ICTs FOR TRANSFER OF TECHNOLOGIES

NIPA® GENX ELECTRONIC RESOURCES & SOLUTIONS P. LTD.
New Delhi-110 034

ICTs FOR TRANSFER OF TECHNOLOGIES

TOOLS & TECHNIQUES

S.R. Verma
(Ph.D., PGDAEM, PGDBA)
Krishi Vigyan Kendra, Bundi
Maharana Pratap University of Agriculture & Technology
Udaipur, Rajasthan

NIPA® GENX ELECTRONIC RESOURCES & SOLUTIONS P. LTD.
New Delhi-110 034

NIPA® GENX ELECTRONIC
RESOURCES & SOLUTIONS P. LTD.

101,103, Vikas Surya Plaza, CU Block
L.S.C. Market, Pitam Pura, New Delhi-110 034
Ph : +91 11 27341616, 27341717, 27341718
E-mail: newindiapublishingagency@gmail.com
web: www.nipabooks.com

For customer assistance, please contact
Phone: + 91-11-27 34 17 17
Fax: + 91-11- 27 34 16 16
E-Mail: feedbacks@nipabooks.com

ISBN: 978-81-19215-32-4

Composed and Designed by NIPA®.

Preface

Agriculture sector is critically important in any developing economy and no less so in India. Where it contribute close to 13.9 percent of GDP and where 60 percent of population is depend on agriculture. Thus even today agriculture is the main occupation for more than half of Indian population. Therefore prosperity of the farmers holds the key of Indian economy. In the present scenario of agriculture new innovations are emerging fast. Intensive use of innovative technologies can improve the economic condition of the farmers. The work of agriculture technology application at farmer's fields is being carried out by state agriculture extension system. Unfortunately agriculture extension systems of majority of states have been failed in achieving farmer's prosperity and happiness which is cause of concern. The foremost grounds of failure of extension system are weak leadership, negligence of bureaucracy, lack of willpower among extension professionals, shortage of budget, lack of skilled human resource, less use of information technology, poor ICT infrastructure, unavailability of knowledge resources and so on. Despite above hindrance some states like M.P., A.P., Kerala, Maharastra and Gujarat have performed excellent work in agriculture development. Enabler of these states is vast use of Information and Communication Technologies for agriculture extension. Every extension professional definitely would like to know the innovative ideas implemented and ICTs used in these states because replication of these ideas and ICTs can revitalize agriculture extension systems of other states. At the present time no book is available which can systematically present the modern ICTs for agriculture extension. I got the idea of writing from here. I believe that it would be great opportunity to the readers to have book containing best ICT initiatives and ideas for agriculture extension. Indeed this book would brainstorm the readers mind and generate unique ideas for agriculture development.

In the entire book efforts has been made to explain the practical application of modern ICTs for transfer of agriculture technologies. I have focused on systematic presentation of various ICT tools with their applications by different organizations. Deep understanding of modern ICT tools and techniques would enable extension workers, scientists, teachers, students and researchers to work efficiently for wellbeing of the farming community. I am sure that through study of this book will provide huge sources of authentic and scientific agriculture information. The book "ICTs for Transfer of Technologies: Tools & Techniques" is useful and informative for officials of government and non-government organizations working for transfer of agriculture technology; teachers and students of agriculture and allied sectors; Officials of Krishi Vigyan Kendras and agriculture universities; Scientist and researchers of agriculture extension and rural development. Overall the book is constructive for Teaching, Research and Extension in the agriculture and allied fields.

I offer my gratitude and sincere thanks to Dr. F.L. Sharma, Head, Department of Extension Education, Rajasthan College of Agriculture, MPUAT, Udaipur for their guidance and continuous motivation to write this book. I am also thankful to chapter contributors for their cooperation.

Although the words are short and few to recognize the sacrifice, love, enormous affection, long patience and cooperation of my wife Mrs. Saroj Verma and my children Prashant & Ayush in completion of this Book. I am really grateful to my family.

Date: 24 March 2015 **S.R. Verma**

Contents

CHAPTER 1

Information and Communication Technology – An Overview

The present era is rightly termed as an "Information era". Therefore, information and wide access to it is considered as wealth. People want adequate and authentic information as early as possible. In recent years, information delivery system has been changed from old ways to the modern ways. At the present time information and communication technologies (ICTs) are increasingly being used as effective tool for reaching and delivering professional services among urban and rural communities. Yet the benefits of the information revolution are still much debated, particularly, in case of developing countries like India. There is a growing concern about the gap between the information "haves and have not's", which will continue to grow unless the developing countries acquire the infrastructure and resources to access these new ICTs. In the past few decades, information and communication technologies have provided society with a vast array of new communication capabilities. For example, people can communicate in real-time with others in different countries using technologies such as instant messaging, voice over IP (VoIP), and video-conferencing. Social networking websites like facebook allow users from all over the world to remain in contact and communicate on a regular basis. Thus modern information and communication technologies have created a "global village," in which people can communicate with each others across the world. Many countries around the world have established organizations for the promotion of ICTs, because it is feared that unless less technologically advanced areas have a chance to catch up, the increasing technological advances in developed nations will only serve to exacerbate the already existing economic gap between technological "have" and "have not" areas. Internationally, the United Nations actively promotes ICTs for Development (ICT4D) as a means of bridging the digital divide.

Meaning and Definitions

The word 'ICT' stands for "Information and Communication Technology". ICT refers to technologies that provide access to information through telecommunications. It is similar to Information Technology (IT), but focuses primarily on communication technologies. This includes the Internet, wireless networks, cell phones, and other communication mediums. The term ICT is sometimes used in preference to Information Technology (IT), particularly in two communities; education and government. In the common usage it is often assumed that ICT is synonymous with IT.

The definition of ICT needs a serious consideration of the general confusion in the understanding of the word 'Technology'. The technology is the process of using scientific material and human resources in order to meet human need or purpose. If we consider a simple definition of Information as 'which can be communicated and understood', we can then put together a basic definition of IT as: Information Technology is the use of information in order to meet human need or purpose. The definition of ICT therefore became: The use of information in order to meet human need or purpose including reference to the use of contemporary devices such as the computer and Internet.

Information and Communications Technology (ICT) is a global term that includes all technologies for the manipulation and communication of information encompassing: computers, internet, cell phones, network hardware & software, satellite systems and so on, as well as the various services and applications associated with them. ICTs in fact encompass any medium for recording and broadcasting of information.

Information and communication technologies are defined as a diverse set of technological tools and resources used to transmit, store, create, share or exchange information. These technological tools and resources include computers, the Internet (websites, blogs and e-mails), live broadcasting technologies (radio, television and webcasting), recorded broadcasting technologies (podcasting, audio and video players, and storage devices) and telephony (fixed or mobile, satellite, video-conferencing etc.).

In the other words ICTs can be defined as electronic and digital technologies for storing, processing, transferring of information and communication. They are enabling technologies that allow quicker and more efficient exchange and processing of information. These new technologies are based on the silicon chips, the laser, fiber optics and a set of varying and diversified technologies. These include a wide variety of computing hardware (PCs, servers, mainframes, networked storage), personal hardware comprising mobile phones, personal devices, MP_3 players, and much more; the full gamut of application software ranges from the smallest home developed spreadsheet to the largest enterprise packages and online software services; and the hardware and software needed to operate networks for transmission of information, again range from a home network to the largest global private networks operated by major commercial enterprises and of course, the internet.

ICT is an umbrella term that includes any communication device or application, encompassing: radio, television, cellular phones, computer and network hardware and software, satellite systems and so on, as well as the various services & applications associated with them, such as video-conferencing and distance learning. ICTs are often spoken of in a particular context, such as ICTs in education, health care and libraries.

Importance and Capabilities of ICT

The positive impact of ICTs is most visible in service sector, where the efficiency levels have gone very high. New businesses like "Business Process Out-sourcing (BPOs)", Banking and Insurance, entertainment, media, communication and other industries as well as organizations, are all taking maximum advantage of the ICT revolution. The importance of ICTs lies less in the technology itself than in its ability to create greater access to information and communication for unreached areas and communities. As far as agriculture sector is concerned appropriate ICT interventions are yielding positive results in developing economies. The "Grameen Phone" initiative in Bangladesh, Kothamale Radio Project in Srilanka, and ITC's e-chaupals in India, are examples of such innovations. The relevance of ICTs for agricultural development in general and for agricultural extension in particular is extremely high for a country like India. During last ten years of its operations, the Kishan Call Centers (KCC) helpline 1800-180-1551 has registered over 4.5 million calls. Internet supporting Information Kiosks are also serving the farming community in many parts of the country. Hence, ICTs are highly relevant to agricultural extension scientists, educationists, researchers, field functionaries and organisations. ICTs are thus emerging as very important tools for agricultural development and it is now essential for every agriculturist to have working knowledge of computers, communications devices, internet, e-mails, expert systems, touch screen systems, information kiosks, mobile telephony, World Wide Web etc.

Modern information and communication technologies have the potential to bypass several stages and sequences of development process. In earlier decades these technologies have proved their potentialities in term of accuracy, cost effectiveness, speed, quality, quantity and timeliness. It may be more appropriate to mention that new communication technologies serve as an aid for extension system not as a replacement of the human element. Over the last two decades, remarkable developments have taken place in information and communication technology. The ICTs like desktop and laptop computers, tablet PCs, internet enabled technologies like e-mail, e-commerce, e-learning, e-conferencing, management information system (MIS), and other online services, cell phones, smart phones and mobile enabled services like SMS, MMS, GPRS, Mass messaging, Interactive voice response services (IVRS), multimedia devices like web camera, digital camera, handy cam, data card, blue-tooth, pen drive, CD-ROMs & DVDs, various types of call centers, information kiosks, touch screen systems, expert systems, teleconferencing, video conferencing, computer assessed services, print media, electronic media, wireless application protocol (WAP), points of presence (pops), FM radio, community radio, web based geographical information system (GIS), Global Positioning System (GPS), and remote sensing have become fairly common among people.

ICT Need in Changing Agricultural Scenario

1. Estimate shows that more than 60 percent of farmers have no access to any source of improved agriculture technologies. Reaching to these unreached farmers is a big challenge before the extension system. ICTs tools like Mobile phone, Internet, Computer, Call centers, Information kiosks, Web portals etc. have potential to address this challenge. ICTs can effectively serve latest agriculture technologies to these unserved farmers.

2. Due to less attention of state governments on agriculture the number of extension personnel are decreasing resulted wide ratio of farmers to extension professional (1000:1). This situation badly affected the extension services. Physically it is very hard to contact every farmer on regular basis but virtually with the help of ICTs every targeted farmers can be served in a better way thus farmer to extension worker ratio can be narrowed and frequency of delivery of extension services can be increased.

3. Right information, at the right time, to the right person, at right place with cost effectiveness is the key factors for decision making in agriculture. Application of ICTs in agriculture can enable extension personnel as well as farmers to perform quick and right decision making.

4. A large number of agriculture technologies are laying at research stations, on the other hand farmers are waiting for new agricultural technologies. Thus a large information gap has been existed between technology generation and their adoption. This gap among the Researchers-Extension

professionals- farmers can be effectively bridge up with application ICT in agriculture.

5. The process of technology transfer in agriculture require sharing of improved knowledge and developing appropriate skills among farmers as well as extension professionals at every step. Inclusion of ICTs in transfer of technology process can achieve desired result.

6. It is proven fact that most of the agriculture operations right from sowing to harvesting are performed by the farm women but due to poor information access they do not involved in the decision making process. ICTs provide equal opportunity to access knowledge resources to the farm women and contribute to right decisions thus ensure gender equality and empower women for participating in the decision making process of the household.

7. Feedback is very important element in the communication process and transfer of technology process also. Many times due to communication barriers the ultimate user of the technology receives distorted information which is misinterpreted, has no meaning to the users. If we have efficient and strong feedback system we can achieve more success in the technology transfer process. ICTs have potential to get fast and accurate feedback about the programme, activity, technology applied or information provided.

8. Majority of Indian farmers have small size of land holdings. Mostly they are avoided by the extension system resulting low productivity. ICTs provide equal opportunity to this majority to come in the mainstream of agriculture development.

9. Majority of Indian farmers have no or little access to the everyday marketing information and sale their produce locally without considering prices in the nearby markets thus fetches low price for their produce. Use of modern ICTs tools can provide everyday marketing information to the farmers and can support them to get better prices for their farm produce.

10. Strong linkages and coordination between Research-Extension-Farmers system can be established by the application of ICTs in agriculture. To achieve faster growth rate in agriculture sector ICTs application is most needed.

11. ICTs can play a crucial role in bridging up the digital divide and technology gap existed among the farming communities.

12. Application of ICTs in agriculture can provide 24 x 7 access to the agriculture technologies.
13. An efficient, strong and prompt monitoring mechanism for agriculture extension programme or schemes is possible through ICTs application.
14. Providing general agriculture information to all farmers is a major weakness of the extension system. This weakness can be converted in to strength with the help of ICTs by providing need based personalized and customized agriculture technology at the door step of the farmer.
15. To develop cost effective information delivery system and qualitative human resource in agriculture we need to promote ICT application in agriculture.

Challenges for Agriculture Extension System

The Indian agricultural sector has been characterized by low productivity growth despite periods of strong growth in the past. To achieve faster productivity growth serious challenges like ICT infrastructure constraints, Agro-input supply chain inefficiencies, wider ratio of farmers to extension worker, digital divide existed among the research-extension & farmers, agriculture marketing deficiencies, shortage of finance and significant problems in the diffusion of and access to agriculture information must be addressed. The challenge for the government and policy makers is to regain agricultural dynamism. To achieve a higher agricultural growth rate, ICT revolution in India must be preceded by the next generation of technology and infrastructure development. Small and marginal farmers, who form the vast majority of Indian farmers, are often unable to access scientific information; they also need to be served.

The National Policy for farmers emphasizes use of Information and Communication Technology (ICT) at village level for reaching out to the farmers with the correct advisories and requisite information. In India about 760 millions of farmers are engaged in agriculture. According to an estimate more than 60 per cent of the farmers have no access to any source of modern agriculture information. Reaching out to these farmers is a great big challenge for the extension system.

Improvement and strengthening of agricultural infrastructure is needed at all levels of the information delivery system, supply chain, input delivery, credit and loans, minimizing post-harvest losses, cold storage chains, marketing *etc.* Shrinking extension is another component of infrastructure that needs attention. The government has a huge research and development infrastructure in the form of institutions such as the Indian Council of Agricultural Research (ICAR),

State Agricultural Universities (SAUs) and Krishi Vigyan Kendras (KVKs) and state agriculture departments. The role of this set-up in education, research and extension activity is of great importance. However, crumbling public agriculture extension system is a cause for concern.

After the green revolution in the mid-sixties, there has been no major technological innovation, which could give a fresh impetus to agricultural productivity. Insufficient extension services and poor access to agriculture information further widen the gap between technology generation and technology adoption which lead to poor productivity levels. A push towards higher agricultural productivity will require an information based decision making agricultural system (precision agriculture). This is often described as the next great revolutionary step in agriculture. Precision agriculture, in turn, is heavily dependent on efficient information dissemination system; information and communication technologies offer the means to set up such a system. The increasing application of ICTs and ICT enabled information services in rural India can reduce information asymmetry and complement the role of extension services.

CHAPTER 2

Modern ICT Tools and Techniques

A tool is a handheld item that aids in accomplishing a task. It is an instrument or machine that aids in achieving a task. Tools are something you use to do a job. ICT tools that can be used for betterment of Agriculture Research, Education and Extension are discussed here.

1. **Computer:** Computer is an advanced electronic device that takes raw data as input from the users and processes these data under the control of set of instructions (called progamme) and gives the result (output) and saves output for the further use. It can process both numerical and non-numerical (arithmetic and logical) calculations.

   Over the past two decades computers have played an important role in the area of agriculture research, education, Extension and management; with the decreasing cost of computer hardware it has become possible to utilize computer technology in agriculture also. At the present time most of the extension personnel are using computer to perform their job better.

2. **Laptops:** A mobile computer that is operated with a battery away from power sources. Newer versions are now wireless and can connect to the Internet in wireless hotspots. Laptop is more useful for agriculture extension services among the farming community. With the help of laptop extension personnel can carry any agriculture technology in the form of video, audio, image, text etc. to show the strength of the technology to the farmers.

3. **Tablet PCs:** Tablet PC is a small screen (12") laptop PC. It is a portable gadget in which data may be directly entered into the screen with a special pen. It's a mini laptop which is more comfortable to carry during the travels.

4. **Smart phone:** A Smartphone is a mobile phone with more advanced computing capability and connectivity than simple cell phone. A typical Smartphone has highly advanced features like relatively large and high resolution touch screen display, Wi-Fi connectivity, web browsing capabilities, and the ability to accept sophisticated applications and operating system capable of running general purpose applications. The majority of these devices run on any of these popular mobile operating systems i.e. Android, Symbian, iOS, BlackBerry OS and Windows Mobile.

5. **Cell phone:** A mobile phone often referred to as a cellular phone or mobile phone is an electronic telecommunication device that can make and receive telephone calls through radio wave or satellite transmission whilst moving around a wide geographic area. Most mobile phones provide voice communications, Short Message Service (SMS), Multimedia Message Service (MMS), short range wireless communication (Infrared, Bluetooth, Wi-Fi), business applications, gaming, photography and internet services such as web browsing and e-mail. Mobile communication technology has become the world's most common way of transmitting voice, data and services, and no other technology has ever spread so fast. The mobile phone is especially important for India because it is growing here faster. Mobile is a very important tool for delivering agriculture extension services among the farmers. Because most of the farmers have mobile phone therefore, it can be used as channel for transmitting agriculture technologies to the farmers at their door step.

6. **Kisan Call Centre:** With the increasing demand of information related to inputs, pesticides, herbicides, high yielding varieties, a farmer today requires guidance of agriculture expert more than anything else. Therefore, the Department of Agriculture & Cooperative (DAC), Ministry of Agriculture, Govt. of India took pioneering initiative and launched Kisan Call Centre (KCC) on January 21, 2004 across the country to deliver extension services to the farming community. The kisan call centre is a toll free telephonic service to the farmers. Any farmer can make call on 1800-180-1551 for any agriculture query on free of cost. Farmers will be answered for their question in his local language. The KCC is a popular service being used by the farmers for getting agriculture expert advices. The KCC services can be extended to the zonal level for providing more customized services to the farmers.

7. **Information Kiosk:** In a generalized term information kiosks are the public installations wherein computers are installed to make agriculture extension services accessible to the people. Information kiosk is a hub of information as per the need of the area or the best source of information.

Any visitor to the kiosks can have access to any kind of information regarding package of practices, plant protection measures, nutritional deficiencies, symptoms of various pests and diseases of several crops and problems he encounters in the field. Even video clippings along with voice can also be viewed and information about technological applications can be listened in local language. Information kiosks are popular ICT tool among the rural and urban areas for getting G2C and B2C services. People are using information kiosks for different services but most of the information kiosks are providing non-agricultural services except some specialized projects. Therefore, it is need of the hour that information kiosks should concentrate to Agriculture extension services so that farming community can be benefitted.

8. **Interactive Whiteboards:** It is an electronic version of a try erase board that enables learners in a virtual class room view, what a instructor, presenter fellow, learner writes or draws. It is also called smart board or electronic whiteboard. Whiteboard surface displays digital files from a computer via a data projector. May function as a standard whiteboard i.e., teacher or student may write on it and then digitize the marked up material. This technology requires a computer, a projector, software and the whiteboard. The computer is connected to the projector displays the computer screen image on the board. It needs to be connected to a pc with data projector attached. Any software applications which are installed on the connected pc can be used on the whiteboard. The whiteboard is controlled with an electronic pen which acts both as a mouse and board pen. The interactive white board is an effective tool to increase learner's interest during the class. It can be used as extension tool for farmers training as well as educational tool for teaching.

Benefits of Using Whiteboard

- Just like conventional boards that the teacher uses, these whiteboards too can help him to explain, clarify a point or draw symbolic figures, diagrams to communicate more clearly.
- They help in broadening the use of e-learning because they demonstrate the potential of alternative modes of delivery.
- They make it easy for the teacher to enhance the presentation content by easily integrating a wide range of material into a lesson, such as, pictures from the internet, a graph or text from files etc.
- They allow learner to participate in group discussion.

9. **Internet:** The internet is a global system of interconnected computer networks that use the standard internet protocol suite (often called TCP/IP) to serve billions of users worldwide. It is a network of networks that consists of millions of private, public, academic, business and government networks of local to global scope, that are linked by a broad array of electronic, wireless and optical networking technologies. The internet carries an extensive range of information resources and services, such as the interlinked hypertext, documents of the World Wide Web (www) and the infrastructure to support e-mails. Internet is being widely used in India. Most of the government, private offices and individuals have internet connectivity. But the use of internet for delivering agriculture extension services to the farmer is not sufficient. The village level extension workers who has internet either personal or official are using internet for personal purposes only. Many of the web portals are providing facility to send SMS to the farmer which can be accessed through internet. A large number of websites and web portals are available for updating professional knowledge of extension workers but only few VLWs are using internet for transfer of technology purpose whereas, Internet has vast capability to speed up extension services in the rural area.

10. **Websites:** A website is a collection of web pages, images, videos or other digital assets that is hosted on one or several web servers, usually accessible via the Internet, cell phone or a Local Area Network (LAN). The three primary things to make a website are domain name, web host and a topic. A website of a company or organisation is a collection of all the facts and information about the organisation in different pages that are included under a domain name.

11. **Webpage:** A webpage is a document, typically written in plain text interspersed with formatting instructions of Hypertext Markup Language (HTML, XHTML). A webpage may incorporate elements from other websites with suitable markup anchors. Web pages are accessed and transported with the Hypertext Transfer Protocol (HTTP) which may optionally employ encryption (HTTP Secure) to provide security and privacy for the user of the webpage content. The user's application, often a web browser, renders the page content according to its HTML markup instructions onto a display terminal.

12. **Web Portal:** A web portal or public portal refers to a website or service that offers a broad array of resources and services, such as e-mail, forums, search engines, and online shopping. Now most of the traditional search engines have transformed themselves into web portals to attract and keep

a larger audience. Portals are used to describe a website that acts as an entry point or gateway to an array of services. It is obvious that all sites do this to a certain degree but a portal has a wide range of resources including search facility, directory of other sites, news, e-mail and lot more. For instance, Yahoo is a great example of a general web portal. There can be industry specific or occupation specific portals like Agriculture portals.

13. **Digital TV:** Similar to analogue TV but has the capacity to deliver rich multimedia learning experiences. It enables interactivity.

14. **Scanners:** Enable the digitization of analogue content. Digital items can then be manipulated by software on the computer and stored.

15. **Personal Digital Entertainment Devices (PDEs) and MP3 Players:** Enable user to download, store and play audio, photo and video files and in many cases, to take part in interactive activities.

16. **Personal Digital Assistants (PDAs):** PDAs and Pocket PCs are also known as Handheld Devices allow input of data via a mini keyboard or equivalent, they usually include a calendar, organizer functions, basic software functions such as word processing, email, spreadsheets, data storage and wireless capacity.

Techniques: Techniques are procedures used to accomplish specific activity or task. There are systematic procedures, formulae, or routine by which task is accomplished. Techniques or technologies supported by internet or mobile phone & computer are described here.

1. **E-mail:** The concept of sending electronic text messages between parties in a way analogous to mailing letters or memos predates the creation of the Internet. Even today it can be important to distinguish between Internet and internal e-mail systems. Internet e-mail may travel and be stored unencrypted on many other networks and machines out of both the sender's and the recipient's control. During this time it is quite possible for the content to be read and even tampered with by third parties, if anyone considers it important enough. Purely internal or intranet mail systems, where the information never leaves the corporate or organization's network, are much more secure, although in any organization there will be IT and other personnel whose job may involve monitoring, and occasionally accessing, the e-mail of other employees not addressed to them. Today you can send pictures and attach files on e-mail. Most e-mail servers today also feature the ability to send e-mail to multiple e-mail addresses.

2. **The World Wide Web:** Many people use the terms Internet and World Wide Web (Web) interchangeably, whereas, these two terms are not synonymous. The World Wide Web is a huge set of interlinked documents, images and other resources, linked by hyperlinks and URLs. These hyperlinks and URLs allow the web servers and other machines that store originals, and cached copies of, these resources to deliver them as required using HTTP (Hypertext Transfer Protocol). HTTP is only one of the communication protocols used on the Internet. In normal use, web browsers, such as Internet Explorer, Firefox, Google chrome, and Apple Safari, access web pages and allow users to navigate from one to another via hyperlinks. Web documents may contain almost any combination of computer data including graphics, sounds, text, video, multimedia and interactive content including games, office applications and scientific demonstrations.
3. **Remote Accessing Technology:** The Internet allows computer users to connect to other computers and information stores easily, wherever they may be across the world. They may do this with or without the use of security, authentication and encryption technologies, depending on the requirements. This is encouraging new ways of working from home, collaboration and information sharing in many industries. An accountant sitting at home can audit the books of a company based in another country, on a server situated in a third country that is remotely maintained by IT specialists in a fourth. These accounts could have been created by home-working bookkeepers, in other remote locations, based on information e-mailed to them from offices all over the world. Remote accessing gives the workers complete access to all of his or her normal files and data, including e-mail and other applications, while away from the office. This concept is also referred to by some network security people as the Virtual Private network, because it extends the secure perimeter of a corporate network into its employees' homes. Application of remote access technology for agriculture may be useful therefore; possibility of application needs to be explored.
4. **Internet Telephony (VoIP):** VoIP stands for Voice-over-Internet Protocol, referring to the protocol that underlies all Internet communication. The idea began in the early 1990s with walkie-talkie-like voice applications for personal computers. In recent years many VoIP systems have become easy to use and as convenient as a normal telephone. The benefit is that, as the Internet carries the voice traffic, VoIP can be free or cost much less than a traditional telephone call, especially over long distances. VoIP is maturing into a competitive

alternative to traditional telephone service. Interoperability between different providers has improved the ability to call or receive a call from a traditional telephone. Simple, inexpensive VoIP network adapters are available that eliminate the need for a personal computer. Voice quality can still vary from call to call but is often equal to and can even exceed that of traditional calls.

Traditional phones are line-powered and operate during a power failure; VoIP does not do so without a backup power source for the phone equipment and the Internet access devices.

5. **Streaming media:** Many existing radio and television broadcasters provide Internet "feeds" of their live audio and video streams (for example, the BBC). These providers have been joined by a range of pure Internet "broadcasters" who never had on-air licenses. This means that an Internet-connected device, such as a computer or something more specific, can be used to access on-line media in much the same way as was previously possible only with a television or radio receiver. Podcasting is a variation on this theme, where usually audio material is downloaded and played back on a computer or shifted to a portable media player to be listened to on the move. These techniques using simple equipment allow anybody, with little censorship or licensing control, to broadcast audio-visual material on a worldwide basis. YouTube is now the leading website for free streaming video with a vast number of users. It uses a flash-based web player to stream and show the video files. Users are able to watch videos without signing up; however, if they do sign up, they are able to upload an unlimited amount of videos and build their own personal profile.

6. **E-learning:** Electronic learning is a type of technology supported education/ learning where the medium of instruction is computer technology, particularly involving digital technologies. E-learning has been defined as "pedagogy empowered by digital technology". In some instances, no in-person interaction takes place. E-learning is used interchangeably in a wide variety of contexts. In companies, it refers to the strategies that use the company network to deliver training courses to employees. In the USA, it is defined as a planned teaching/ learning experience that uses a wide spectrum of technologies, mainly Internet or computer-based, to reach learners. Lately in most universities, e-learning is used to define a specific mode to attend a course or programmes of study where the students rarely, if ever, attend face-to-face for on-campus access to educational facilities, because they study online.

7. **Teleconferencing:** It is interactive group communication between two or more people in two or more locations through an electronic medium. It brings people together under one roof even though they are separated by hundreds of miles. It was first introduced in 1960's with American Telephone. Three type of conferencing are:

   i. Video conferencing: Television like communication augmented with sound.

   ii. Computer conferencing: Printed communication through key board terminals.

   iii. Audio conferencing: Verbal communication via the telephone.

   - **Video conferencing:** Videoconferencing is also known as video teleconferencing. It is a set of interactive telecommunication technologies which allow two more locations to interact in a two way video and audio transmissions simultaneously. It is a remote meeting between two or more individuals present in geographically dispersed locations. It is an effect of virtual presence where a person at a distance sight appears to be in the same room. Users can meet and share information virtually from wherever they are using videoconferencing equipment and communicate using any network available with them.

Advantages of Video Conferencing

- A live conversation between two partners from different locations is possible with the visibility.
- The experts (scientists) can have virtual contact with the farmers and solve the field problems.
- Useful in giving training sessions for guest lectures.
- Useful in monitoring the progress of various activities.
- Reduces the cost of travel and time.
- Researcher collaborates with colleagues at other institutions on a regular basis without loss of time due to travel.
- Problem solving information can be exchanged and procedural tasks can be discussed
- Follow-up for earlier meetings can be done with relative ease and little expense.

8. **GPRS:** GPRS stands for General Packet Radio Service. It is a packet based wireless communication service that provides data rates from 56 to 114 kbps and continuous connection to the Internet for mobile phone and computer users. GPRS is a second generation (2G) and third generation (3G) or sometimes referred to as in-between both generations, 2.5G wireless data service that extends GSM data capabilities for Internet access, multimedia messaging services, and early mobile Internet applications via the wireless application protocol (WAP), as well as other wireless data services. GPRS is a very useful service for the rural India. It provides connectivity and access to the farming communities for agriculture extension services.

9. **Interactive Voice Response Service:** IVRS is a technology that allows a computer system to interact with humans through the use of voice and DTMF tones input via keypad. IVRS technology automates interactions with telephone callers. The system responds with pre recorded or dynamically generated audios to further direct users on how to proceed. IVRS applications can be used to control almost any function where the interface can be broken down into a series of simple interactions. Many organisations are turning to IVRS to reduce the cost of common sales, service, collections, inquiry and support calls to and from their company. In telecommunications, IVRS allows customers to interact with a company's host system via a telephone keypad or by speech recognition, after which they can service their own inquiries by following the IVRS dialogue. Implementation of Interactive voice response service for agriculture extension services can be an innovative step for wellbeing of farmers. Most of the quarries raised by farmers are common which can be addressed by IVRS and for more complex questions option can be given for direct interaction with agriculture specialist.

10. **GIS Technology:** Geographic Information System integrates hardware, software, and data for capturing, managing, analyzing, and displaying all forms of geographically referenced information.GIS allows us to view, understand, question, interpret, and visualize data in many ways that reveal relationships, patterns, and trends in the form of maps, globes, reports, and charts. The system helps to answer questions and solve problems by looking at your data in a way that is quickly understood and easily shared. GIS technology has become a vital tool for crop management. Geographic data and satellite images about soil conditions helps farmers to be more efficient in segmenting arable land in order to apply differential rates of fertilizer, and forecasting to determine when, where, and what to plant in

what is known as precision agriculture. GIS have many applications in agriculture research also.

11. **USSD Technology:** Unstructured Supplementary Service Data (USSD) is a Global System for Mobile communication technology that is used to send text between a mobile phone and an application program in the network. USSD is a protocol that allows phones to communicate with the service provider's computers. It allows high speed interactive communication between the subscribers and applications. It is similar to SMS in the way that it allows to send and receive short text messages. It is different in the way that it is session oriented. Since USSD is session oriented, it gives a very short delay between sending the query and receiving the response. This makes it ideal to query information from the network and to provide text content as a service. An USSD gateway is a platform that enables operators to introduce messaging services with USSD as the bearer, enabling faster response times. The platform is complementary to SMS service and an SMS gateway. An USSD gateway acts as a gateway between applications and the GSM network and lets you deliver USSD messages of up to 182 characters on a network between mobile stations and applications. While SMS provides a store and forward functionality, USSD is not like this. It is a real time session-oriented service, which means that when a user accesses a service with USSD, a session is established and the radio connection stays open until the user, application, or time out releases it. This provides faster response times for interactive applications. An USSD session needs to be allocated to each and every interaction. A typical USSD gateway uses the same application programming interface that the Short Message Service centre does. Therefore, it is easy to port services based on SMS to utilize USSD as the bearer. In practice, only USSD specific modifications are needed to external applications. USSD technology can be used for proving crop based information to the farmers.

12. **EDGE:** Stands for Enhanced Data rates for GSM Evolution. A faster version of GSM wireless service. It enables data to be delivered at rates up to 384 kbps via broadband.

13. **Instant Messaging:** Instant Messaging is a technology that allows users to communicate in real-time, via text chat. Essentially IM system alerts you when a contact on your list is online and you can then initiate a chat session.

14. **MMS:** Stands for Multimedia Message Service. An extension of SMS that allows pictures, sounds or low quality videos to be sent on a wireless network.

15. **SMS:** Stands for Short Messaging Service. SMSs are text messages that can be sent to mobile phones from the Internet or from other mobile devices.

16. **Wi-Fi:** It is wireless internet accessing technology enable the devices to automatically connect with internet in the defined locality. It is standard protocol for networking and connecting to the Internet.

17. **Conference Calling:** Conference calling lets you add several people to a mobile phone call. You can connect the other callers at the start of your call, or add people to a call that's already in progress. Conferencing is made when the person making a call wants to have more than one person listening in to the audio or the talking part of the phone call.

18. **Podcasts:** Podcasting is a method of publishing audio files via the Internet, allowing users to subscribe to a feed to receive new files automatically.

19. **Vodcasts:** Video on demand is the same concept as podcasting but with video files.

20. **Blogs:** A web-based journal or log book. Logs are chronologically ordered web-postings by an author or group of authors. They may be personal, individual records, group collaborations or representative of an institution.

21. **Wikis:** A wiki is a website (or other hypertext document collection) that allows users to add content, on an Internet forum, but also allows anyone to edit the content. "Wiki" also refers the collaborative software used to create such a website.

CHAPTER 3

Type of Networks

What is Computer Network?

A computer network is a collection of computers and devices connected to each other. The network allows computers to communicate with each other and share resources and information. The Advance Research Projects Agency (ARPA) designed "Advanced Research Projects Agency Network" (ARPANET) for the United States Department of Defence. It was the first computer network in the world in late 1960's and early 1970's

Why Networking: The following are the important needs for computer networking.

- **File sharing:** Networking of computers helps the users to share data files. Computers connected to a network can share files and documents with each other. Personal computers connected to a business network can choose which files and folders are available to share on the network.
- **Hardware sharing:** Users can share devices such as printers, scanners, CD-ROM drives, hard drives etc. A computer can print files stores on a computer with a printer connected on the network. Additionally, printers can be connected using a print server, which allows direct printing from all computers.
- **Application sharing:** Applications can be shared over the network, and this allows implementing client/server applications.
- **User communication:** Networks allow users to communicate using e-mail, newsgroups, and video conferencing etc.

- **Network gaming:** A lot of network games are available, which allow multi-users to play from different locations.
- **Sharing Media:** Sharing media between computers is easy when connected to a network like file sharing. Computers can stream music, videos and movies from one computer to the next.

Classification of Networks on the Basis of Scale Method

Based on their scale, networks can be classified as Local Area Network (LAN), Wide Area Network (WAN), Metropolitan Area Network (MAN), Personal Area Network (PAN), Virtual Private Network (VPN), Campus Area Network (CAN), Storage Area Network (SAN) etc. The most common computer networks are given below.

1. **Personal Area Network (PAN):** A personal area network is a computer network used for communication among computer devices close to one person. Some examples of devices that are used in a PAN are printers, fax machines, telephones, PDAs and scanners. The reach of a PAN is typically about 20-30 feet (approximately 6-9 meters), but this is expected to increase with technology improvements.
2. **Local Area Network (LAN):** A local area network is a computer network covering a small physical area, like a home, office, or small group of buildings, such as a school, or an airport. Current wired LANs are most likely to be based on Ethernet technology, although new standards like ITU-T G.hn also provide a way to create a wired LAN using existing home wires (coaxial cables, phone lines and power lines). For example, a library may have a wired or wireless LAN for users to interconnect local devices (e.g., printers and servers) and to connect to the internet. All interconnected devices must understand the network layer (layer 3), because they are handling multiple subnets. The defining characteristics of LANs, in contrast to WANs (wide area networks), include their higher data transfer rates, smaller geographic range, and lack of a need for leased telecommunication lines. Current Ethernet or other IEEE 802.3 LAN technologies operate at speeds up to 10 Gbit/s. This is the data transfer rate. IEEE has projects investigating the standardization of 100 Gbit/s, and possibly 400 Gbit/s.
3. **Campus Area Network (CAN):** A campus area network is a computer network made up of an interconnection of local area networks (LANs) within a limited geographical area. It can be considered one form of a metropolitan area network, specific to an academic setting. In the case of

a university campus based, campus area network, the network is likely to link a variety of campus buildings including; academic departments, the university library and student residence halls. A campus area network is larger than a local area network but smaller than a wide area network (WAN). The main aim of a campus area network is to facilitate students accessing internet and university resources. This term is most often used to discuss the implementation of networks for a contiguous area. This should not be confused with a Controller Area Network.

4. **Metropolitan Area Network (MAN):** A metropolitan area network is a network that connects two or more local area networks or campus area networks together but does not extend beyond the boundaries of the immediate town/city. Routers, switches and hubs are connected to create a metropolitan area network.

5. **Wide Area Network (WAN):** A wide area network is a computer network that covers a broad area (i.e. any network whose communications links cross metropolitan, regional, or national boundaries. Less formally, a WAN is a network that uses routers and public communications links. Contrast with personal area networks (PANs), local area networks (LANs), campus area networks (CANs), or metropolitan area networks (MANs), which are usually limited to a room, building, campus or specific metropolitan area (e.g., a city) respectively. The largest and most well-known example of a WAN is the Internet. A WAN is a data communications network that covers a relatively broad geographic area (i.e. one city to another and one country to another) and that often uses transmission facilities provided by common carriers, such as telephone companies. WAN technologies generally function at the lower three layers of the OSI reference model: the physical layer, the data link layer, and the network layer.

6. **Global Area Network (GAN):** A global area network specification is in development by several groups, and there is no common definition. In general, however, a GAN is a model for supporting mobile communications across an arbitrary number of wireless LANs, satellite coverage areas etc. The key challenge in mobile communications is "handing off" the user communications from one local coverage area to the next. In IEEE Project 802, this involves a succession of terrestrial WIRELESS local area networks (WLAN).

7. **Virtual Private Network (VPN):** A virtual private network is a computer network in which some of the links between nodes are carried by open connections or virtual circuits in some larger network (e.g., the Internet) instead of by physical wires. The link-layer protocols of the virtual network

are said to be tunnelled through the larger network when this is the case. One common application is secure communications through the public Internet, but a VPN need not have explicit security features, such as authentication or content encryption. VPNs, for example, can be used to separate the traffic of different user communities over an underlying network with strong security features. A VPN may have best-effort performance, or may have a defined service level agreement (SLA) between the VPN customer and the VPN service provider. Generally, a VPN has a topology more complex than point-to-point. A VPN allows computer users to appear to be editing from an IP address location other than the one which connects the actual computer to the Internet.

Classification of Networks on the Basis of Connection Method

Computer networks can also be classified according to the hardware and software technology that is used to interconnect the individual devices in the network, such as Optical fiber, Ethernet, Wireless LAN, Home PNA, Power line communication or G.hn.

1. **Ethernet:** Ethernet uses physical wiring to connect devices. Frequently deployed devices include hubs, switches, bridges and routers.
2. **Wireless LAN:** Wireless technology is designed to connect devices without wiring. These devices use radio waves or infrared signals as a transmission medium.
3. **Power line communication or G.hn:** ITU-T G.hn technology uses existing home wiring (coaxial cable, phone lines and power lines) to create a high-speed (up to 1 Gigabit/s) local area network.

Networking of Networks (Internet)

Internetworking involves connecting two or more distinct computer networks or network segments via a common routing technology. The result is called an internetwork (often shortened to Internet). Two or more networks or network segments connected using devices that operate at layer 3 (the 'network' layer) of the OSI Basic Reference Model, such as a router. Any interconnection among or between public, private, commercial, industrial, or governmental networks may also be defined as an internetwork. In modern practice, the interconnected networks use the Internet Protocol. There are at least three variants of internetwork (INTRANET, EXTRANET, INTERNET) depending on who administers and who participates in them:

Intranets and extranets may or may not have connections to the Internet. If connected to the Internet, the intranet or extranet is normally protected from being accessed from the Internet without proper authorization. The Internet is not considered to be a part of the intranet or extranet, although it may serve as a portal for access to portions of an extranet.

a. **INTRANET:** An intranet is a set of networks, using the Internet Protocol and IP-based tools such as web browsers and file transfer applications that are under the control of a single administrative entity. That administrative entity closes the intranet to all but specific, authorized users. Most commonly, an intranet is the internal network of an organization. A large intranet will typically have at least one web server to provide users with organizational information.

b. **EXTRANET:** An extranet is a network or internetwork that is limited in scope to a single organization or entity but which also has limited connections to the networks of one or more other usually, but not necessarily, trusted organizations or entities (e.g. a company's customers may be given access to some part of its intranet creating in this way an extranet, while at the same time the customers may not be considered 'trusted' from a security standpoint). Technically, an extranet may also be categorized as a CAN, MAN, WAN, or other type of network although by definition an extranet cannot consist of a single LAN it must have at least one connection with an external network.

c. **INTERNET:** The Internet is a specific internetwork. It consists of a worldwide interconnection of governmental, academic, public, and private networks based upon the networking technologies of the Internet Protocol Suite. It is the successor of the Advanced Research Projects Agency Network (ARPANET) developed by DARPA of the U.S. Department of Defence. The Internet is also the communications backbone underlying the World Wide Web (WWW). The 'Internet' is most commonly spelled with a capital 'I' as a proper noun, for historical reasons and to distinguish it from other generic internetworks. Participants in the Internet use a diverse array of methods of several hundred documented, and often standardized, protocols compatible with the Internet Protocol Suite and an addressing system (IP Addresses) administered by the Internet Assigned Numbers Authority and address registries. Service providers and large enterprises exchange information about the reach ability of their address spaces through the Border Gateway Protocol (BGP), forming a redundant worldwide mesh of transmission paths.

Hardware Components for Networking

All networks are made up of basic hardware building blocks to interconnect network nodes, such as Network Interface Cards (NICs), Bridges, Hubs, Switches, and Routers. In addition, some methods of connecting these building blocks are required, usually in the form of galvanic cable (most commonly Category 5 cable). Less common are microwave links (as in IEEE 802.12) or optical fibre cable. An Ethernet card may also be required.

1. **Network interface cards:** A network card, network adapter or NIC (Network Interface Card) is a piece of computer hardware designed to allow computers to communicate over a computer network. It provides physical access to a networking medium and often provides a low level addressing system through the use of MAC addresses.
2. **Repeaters:** A repeater is an electronic device that receives a signal and retransmits it at a higher power level or to the other side of an obstruction so that the signal can cover longer distances without degradation. In most twisted pair Ethernet configurations repeaters are required for cable which runs longer than 100 meters.
3. **Hubs:** A hub consiste multiple ports. When a packet arrives at one port, it is copied unmodified to all ports of the hub for transmission. The destination address in the frame is not changed to a broadcast address.
4. **Bridges:** A network bridge connects multiple network segments at the data link layer (layer 2) of the OSI model. Bridges do not promiscuously copy traffic to all ports, as hubs do, but learn which MAC addresses are reachable through specific ports. Once the bridge associates a port and an address, it will send traffic for that address only to that port. Bridges do send broadcasts to all ports except the one on which the broadcast was received. Bridges learn the association of ports and addresses by examining the source address of frames that it sees on various ports. Once a frame arrives through a port, its source address is stored and the bridge assumes that MAC address is associated with that port. The first time that a previously unknown destination address is seen, the bridge will forward the frame to all ports other than the one on which the frame arrived. Bridges come in three basic types: Local bridges: Directly connect local area networks (LANs). Remote bridges: Can be used to create a wide area network (WAN) link between LANs. Remote bridges, where the connecting link is slower than the end networks, largely have been replaced by routers. Wireless bridges: Can be used to join LANs or connect remote stations to LANs.

5. **Switches:** A switch is a device that forwards and filters OSI layer 2 data grams (chunk of data communication) between ports (connected cables) based on the MAC addresses in the packets. Switch is a marketing term that encompasses routers and bridges, as well as devices that may distribute traffic on load or by application content (e.g., a Web URL identifier). Switches may operate at one or more OSI model layers, including physical, data link, network, or transport (i.e., end-to-end). A device that operates simultaneously at more than one of these layers is called a multilayer switch.
6. **Routers:** Routers are networking devices that forward data packets between networks using headers and forwarding tables to determine the best path to forward the packets. Routers work at the network layer.

Advantages of Internetworking

1. **Education:** Networking allow to get additional information by the students, teachers and scientists. It is a medium for interactive and collaborative learning. It is very useful for distance education.
2. **Publishing:** All newspapers and newsletters are available on internet.
3. **Shopping:** E-commerce is possible.
4. **Advertising:** Useful for advertising the products with text, graphics and pictures and video clippings.
5. **Financial services:** Stock broking and research reports on stocks are available and can be downloaded. The transactions like tele credit card checking, tele banking, tele insurance are taking place.
6. **The business of governance:** Public information useful for every citizen can be kept on the net. Government services can be made available and can provide fast, transparent services through Internet. E.g., E –seva in AP.
7. **Career:** Career opportunities can be known with the help of net.
8. **Internet communication:** Provides access to all kinds of information available on the latest technology in any field.
9. **E-mail:** Is the primary communication tool on internet. One can send and receive mails without any geographical barriers. We can send e-mail through websites like rediffmail.com, hotmail.com, gmail.com etc.
10. **Lister:** It allows group of people with common interest to send messages to each other at cost.

11. **Usenet news group:** A newsgroup is worldwide platform for exchanging ideas and information by common minded people. The mail programme is loaded with windows outlook express. It helps use to send, receive and store e-mails.

12. **World Wide Web (www)** is a network of information resources. The digital pages on www are called web sites. The first page of website is called Home page.

CHAPTER 4

Cellular Phone Services and Applications

[1]Krunalkumar C. Kamani and [2]Yogesh R. Ghodasara

[1]Assistant Professor (Computer Science)
[2]Principal & Dean, College of Agricultural Information Technology
AAU, Anand (Gujarat)

A cellular phone is also known as a mobile phone, cell phone, portable phone and a hand phone. It is a device that can make and receive telephone calls over a radio link while moving around a wide geographic area. It does so by connecting to a cellular network provided by a mobile phone operator, allowing access to the public telephone network.

The cell phone was invented by Dr. Martin Cooper who led a team of developers at Motorola and made the first cell phone call in April 3, 1973 at Motorola.

Today, cell phones are becoming capable of doing almost anything a computer is capable to do. Below, are few examples of what a phone can do?

- Take and make calls to any other person with a phone.
- Send and receive text messages. We usually called it SMS (short message service)
- Take, view, and store pictures and videos using embedded cameras.
- Access the Internet, e-mail, chat, and even download add-ons for the phone.
- Play games and have access to common programs such as a calculator, contacts, calendar, clock, to-do list, etc.

- Store and run files.
- Play music.

History of Cell Phone

Inventor Charles E. Alden claimed, in the 29 April 1906 issue of the New York World, to have invented a device called the "vest pocket telephone" although Alden never had the chance to produce this device in large quantities.

The first mobile telephone calls were made from cars in 1946. Bell System's Mobile Telephone Service was made on 17 June in St. Louis, Missouri, followed by Illinois Bell Telephone Company's car radiotelephone service in Chicago on 2 October. The MTA phones were composed of vacuum tubes and relays, and weighed over 80 pounds (36 kg). There were initially only 3 channels for all the users in the metropolitan area, increasing later to 32 channels across 3 bands. This service continued into the 1980s in large portions of North America. Due to the small number of radio frequencies available, the service quickly reached capacity. In 1956, the world's first partly automatic car phone system, Mobile System was introduced in Sweden.

John F. Mitchell, Motorola's chief of portable communication products in 1973, played a key role in advancing the development of handheld mobile telephone equipment. Mitchell successfully pushed Motorola to develop wireless communication products that would be small enough to use anywhere and participated in the design of the cellular phone. Martin Cooper, a Motorola researcher and executive, was the key researcher on Mitchell's team that developed the first hand-held mobile telephone for use on a cellular network. Using a somewhat heavy portable handset, Cooper made the first call on a handheld mobile phone on 3 April 1973 to his rival, Dr. Joel S. Engel of Bell Labs.

Components of Mobile Phone

The following common components are found in all cell phones:

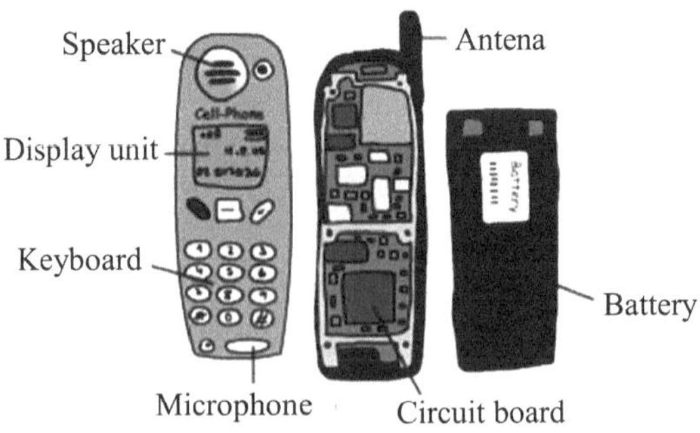

Fig. 1: Components of cell phone

- An amazing circuit board containing the brains of the phone
- An antenna
- A liquid crystal display(LCD)
- An input mechanism to allow the user to interact with the phone. The most common input mechanism is a keypad, but touch screens are also found in some high-end smart phones.
- A microphone
- A speaker
- A battery, providing the power source for the phone functions.
- All GSM (Global System for Mobile communications) phones use a SIM (Subscriber Identification Module) card to allow an account to be swapped among devices. Some CDMA (Code Division Multiple Access) devices also have a similar card called a R-UIM.
- Individual GSM, WCDMA, iDEN and some satellite phone devices are uniquely identified by an International Mobile Equipment Identity (IMEI) number.

Why Mobile Phones are Called Cell Phones

One of the most interesting things about a cell phone is that it is really a radio. Before cell phones, people who needed mobile communications ability installed radio telephones in their cars. In the radio telephone system, there was one central antenna tower per city, and 25 channels available on that tower. The cellular phone system divides the area of a city into small cells. This allows extensive frequency reuse across a city, so that millions of people can use cell phones simultaneously.

In a typical analog cell-phone system in the United States, the cell-phone carrier receives about 800 frequencies to use across the city. The carrier chops up the city into cells. Each cell is typically sized at about 10 square miles (26 square kilometres). Cells are normally thought of as hexagons on a big hexagonal grid, like this:

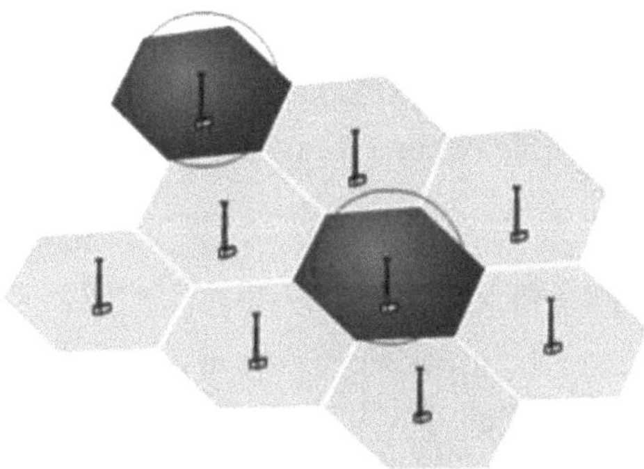

Fig. 2: cells in cellular network

Because cell phones and base stations use low-power transmitters, the same frequencies can be reused in non-adjacent cells. The two black cells can reuse the same frequencies.

How Cell Phone Works

The carrier chops up an area, such as a city, into cells. Each cell is typically sized at about 10 square miles (perhaps 3 miles x 3 miles). Cells are normally thought of as hexagons on a big hexagonal grid. Each cell has a base station that consists of a tower and a small building containing the radio equipment. Cell phones have low-power transmitters in them and the base station is also transmitting at low power. Low-power transmitters have two advantages:

1. The power consumption of the cell phone, which is normally battery-operated, is relatively low. Low power means small batteries, and this is what has made handheld cellular phones possible.
2. The transmissions of a base station and the phones within its cell do not make it very far outside that cell. Therefore, cells can use the same 56 frequencies. The same frequencies can be reused extensively across the city.

The cellular approach requires a large number of base stations in a city of any size. A typical large city can have hundreds of towers. But because so many people are using cell phones, costs remain fairly low per user. Each carrier in each city also runs one central office called the Mobile Telephone Switching Office (MTSO). This office handles all of the phone connections to the normal land-based phone system, and controls all of the base stations in the region.

As you move toward the edge of your cell, your cell's base station will note that your signal strength is diminishing. Meanwhile, the base station in the cell you are moving toward (which is listening and measuring signal strength on all frequencies, not just its own one-seventh) will be able to see your phone's signal strength increasing. The two base stations coordinate themselves through the MTSO, and at some point, your phone gets a signal on a control channel telling it to change frequencies. This hand off switches your phone to the new cell.

Types of Mobile Phone

Due to its utility, necessity and affordability, mobile phone is one of the most ubiquitous gadgets that one finds in the world. Every mobile phone manufacturer comes out with its own range of offerings with something new, something different and something unexpected being always the high point. And with the value added features varying from music/sound capabilities to web browsing abilities dominating user's choice, the voice call function seems to have taken more of a back seat. But not every mobile phone comes loaded with all the features. They all have some underlying utility and other aspects for which they are designed and offered in the market. And accordingly they can be broadly classified into following categories;

Basic Mobile Phone: It is the simplest and the most original of all the mobile phones that you would see in the market. The cheapest of them all, it is sold in maximum numbers. Though with cost of technology falling and customer expectations rising most of the basic mobile phones now come with value additions like ring tone options, etc. Most of the mobile phone brands offer basic models.

Smartphone: These phones are for those who want a single gadget to be fully loaded so that it can serve any and every purpose from communication to computing and from entertainment to productivity. Smartphone do multiple tasks from e-mail and web browsing to playing music, capturing and playing videos, storing and managing data and many other functions too. These have higher price tags but the features they offer fully justify their price. RIM's BlackBerry, Nokia's N97 to Lumia series, Micromax's A111, Samsung's Core and Apple's iPhone are some of the most popular smart phones. Most of the latest mobile phones fall in this category.

PDA Phones: Personal Data Assistant phones are the favourite companion of business executives. Running on operating systems like MS Mobile Pocket PC and Palm OS, these are offered by HP, Palm One. In fact the mobile phone market has become so demanding, dynamic and competitive that the line that divides smart phones from PDA phones is fast fading.

GSM/CDMA Mobile Phones: GSM (Global System for Mobile Communications) and CDMA (Code Division Multiple Access) are the two most important and commonly used networking technologies and your phone would be designed to work on either. Both the technologies have their own pros and cons and same get transferred to the corresponding mobile phone. The latest innovations though have now made it possible to use both the technologies in the same mobile phone.

Others types: Mobile phones can also be categorized as lifestyle phones, business phones, entertainment phones, gaming phones, etc. depending upon the features and consumer market segment that these may be targeting. The best thing is that you can make a mental note of what all you want in your mobile phone and you will be surprised to find that it is already available in the market.

Advantages of Mobile Phone

1. **Stay connected 24 x 7 globally:** The most basic benefit of a cell phone for which most of us use it is that we can stay connected with our loved ones in any part of the world and anytime. Gone are the days when we used to stand in queues to make an STD or ISD calls. You can talk to your loved ones staying even seven seas far with mobile phones.
2. **Short Message Service (SMS):** When initially SMS was invented, the makers were not actually sure whether it will work. I mean when people can straightaway make a call and talk, then why would anybody like to take the pain of typing a message! But to their surprise SMSs are today the most widely used service across the world. There are many situations in which a person can't attend a call, so all you have to do is simply send an SMS and without talking your message is delivered.
3. **Your way out in emergency:** Nothing more than a cell phone comes to great help in emergency. Imagine you are stuck in a traffic jam, getting late for a meeting or your car has broken down in middle of nowhere. Mobile phones are of great use in time of such emergencies. You can contact help with the use of mobile phone easily.
4. **Map reading in your hand:** Mobile phones are constantly being upgraded with new technology and the recent phones are equipped with navigation and GPRS systems. You can never get lost if you have a cell phone with navigation system. Imagine, it gives you the details of every nook and corner of your destination.
5. **Mini PC:** Mobile phones are nowadays almost equivalent to mini computers. The latest ones are equipped with windows and internet facilities. So you don't need to wait for the newspaper! You can simply access the internet on your cell phone and get to know about the latest news, your e-mails, movie shows and a lot more.
6. **Improve your trade:** With mobile you can constantly stay in touch with your employees and get to know about crucial information of your business, make online trading using m-commerce (mobile commerce).
7. **Help in legal issues:** A lot of criminals are these days being held because of their mobile phones. The police can track a criminal via tracking the place where his / her mobile phone is using GPS. Also checking a mobile phone's call records give vital information to the defense forces about the criminals. In other words we can say that it is helpful to solve cyber crime cases.

8. **The great entertainer**: With a mobile phone in your hand, you don't need a TV or PC to get entertained. It is all in your mobile phone. You can play games, listen to music, and click pictures and even record videos in your mobile phone. In short along with phone calls it is used for messaging, video, songs, games, alarm clock, notes, calendar, reminder, internet browsing, navigation, learning etc.

9. **Transfer of data:** Nowadays, mobile phones are equipped with infrared, Wi-Fi and Bluetooth technologies which allow you to transfer data like mails, pictures, music and even videos just in span of seconds. Even one can pay their bill, book tickets and perform financial transaction using mobile phone.

10. **Status and Style:** Mobile phones have become a matter of status or prestige and fashion statement, especially among the youth. If you have a latest handset then you can definitely impress others.

Disadvantages of Mobile Phone

1. **Negative Health Effects:** Mobile phones transmit and receive Radio Frequency (RF) signals in order to communicate. The RF signals from mobile phones fall within the microwave part of the electromagnetic spectrum. This radiation is also referred to as microwave radiation or electromagnetic radiation. The surveys also indicate and suggest that the use of mobile phones by men is associated with a decrease in semen quality. The decrease in sperm count, motility, viability, and normal morphology is related to the duration of exposure to cell phones. Use of hands-free (wired/blue-tooth) can at times pass on loud sounds to our ears which can result in weakening of ear-drums. Nowadays, one can download lots of songs, so keeping the hands-free glued inside your ears for long hours can really affect the sensitivity of ears in the long run of life. It's a bitter fact that, the surface of a cell phone has millions of bacteria and virus on it and that can be a strong reason of immediate skin problem on face or can result into other internal infections wherein the microbes creep inside the body through mouth or other openings.

2. **Driving Safety:** A simple Google search will show hundreds of studies on the subject of mobile phone use and driving. All the studies once can agree that mobile phone use while driving increases the risk of an accident. This increased risk is due to the inability to fully focus on driving while, and shortly after, talking on a mobile phone. Use of the telephone in hands free mode is less risky than holding the telephone to the ear with one hand while talking.

3. **The Effect on Children:** Children that grow up in our new mobile phone world are exposed to microwave radiation from the moment they are in the womb. However, the effects of microwave radiation on infants are unknown to most parents. The Stewart Report is a report commissioned by the British Parliament in response to fears that mobile phone use could be linked to memory loss, and even Alzheimer's disease.
4. **Tracing:** Your SIM can be exploited as tracking device and if you're an important person then that can be a big concern for you.
5. **Disturbance:** It can be a big time distraction and nuisance in calm and silent places like libraries, cinemas, restaurants, etc. Some cell phone users lose the sense of deciding when and where they can talk on the mobile phone and where they can't, without slightest consideration for the fellow beings around.
6. **Unwanted Advertisements:** The mobile phone advertisements through messages are becoming a pain for the cell phone users.
7. **Battery Blast:** There have been cases of mobile phone blasts, due to the excessive heating up of its battery. This can be a fatal issue although rare.
8. **Environment Issues:** Many components of mobile phones are considered toxic like arsenic, lithium, cadmium, copper, lead, mercury and zinc. These poisonous substances may leach from decomposing waste in landfills, seep into groundwater and contaminate the soil. Metals build up in the soil, can then enter the food chain and in sufficient concentrations may cause health problems. But not just the dumping of mobile phones is dangerous for the environment. The production of new mobile phones contributes to climate change by using up energy and virgin materials which release greenhouse gasses into the atmosphere. It is estimated that up to 90 percent of these greenhouse gases can be saved by recycling materials from mobile phones.
9. **Recycle:** In order to overcome with this problem the recycling movement is carried out by many NGOs and mobile phone companies. When your phone has come to the end of its life donate it to recycle and clean up environment and help to move towards go green. We have mentioned to help you use your mobile phone in a responsible manner. We humans created mobile phones and it is all up to our control.

Benefits of Mobile Phone Application to Agriculture

Mobile telephony effectively reduces the distance between individuals and institutions, making the sharing of information and knowledge easier and more effective. The mobile telephone is no longer just an audio communication tool but capable of providing additional integrated functions. Following are the advantages of mobile phone in agriculture.

1. **Help Farmers Raise Their Incomes:** Mobile phones can serve as the backbone for early warning systems to mitigate agricultural risks and safeguard agricultural incomes. Many countries provides, local weather forecasts transmitted through SMS provided very timely warnings of impending frosts or conditions that favored pests.
2. **Making Agricultural Marketing More Efficient:** Mobile phones, in addition to other ICTs, can provide information to both producers and consumers of the prices offered for agricultural products in various locations.
3. **Mobile Phones Revolutionize Livestock Sector:** Farmers and veterinarians across the country are increasingly using cell phones to issue alerts quickly about possible animal disease outbreaks at a very early stage and to track wide-scale vaccination campaigns. Mobile phone applications are making 'early warning' a matter of seconds instead of weeks.
4. **Lowering the Costs of Information:** The most obvious and cross-cutting way that mobile phones can improve agriculture is by improving access to information and making it less costly to obtain. In many rural areas, the arrival of mobile coverage is a radical change in the nature of the information ecosystem.
5. **Mobile Phone Q&A for Farmers:** Nowadays, the mobile phones provide a direct communication to scientist through kisan call center for their question related to agriculture and allied sector. Even with smart phone one can directly perform video conference, offering farmers on their mobile phones expert answers to their agricultural queries, and all delivered by voice rather than text to avoid the literacy and language barrier.
6. **Reduce Transport Costs:** Mobile phones may substitute travel. Where safety standards are minimal, roads are in disrepair and distances are great, substituting phone calls for travel reduces farmers' time and cost burdens. Time savings are important for agricultural households, because many crops have extremely time-sensitive and labor-intensive production cycles. Transportation cannot be avoided entirely: Crops need to get to customers. Although mobiles can inform farmers where they should travel to market their crops, evidence suggests that the wealthy maintain an

advantage in their ability to make use of this information .In combination with improved rural roads, ICT will encourage larger truck-traders to visit harder-to-reach areas, connecting rural and urban regions.

7. **A Platform for Service Delivery and Innovation:** The numerous capabilities of mobile phones provide sufficient opportunities to deliver traditional and innovative services. Traditional agricultural extension agents are increasingly being outfitted with mobile phones through programs to increase their effectiveness by networking them to knowledge banks. Extension can reach more clients through mobile-based learning platforms textual or richer platforms, such as video that provide tips to farmers to improve agricultural skills and knowledge.

8. **Mobile Controlled Water Pump:** Mobile technology has created a big revolution in today's world. We can use this technology for effective communication throughout the world and contact any one without delay. Even we can operate our agricultural water pump sets with this technology. Due to poor & unpredictable availability of Electricity in rural areas, farmers have to start their pumps at many times during the whole day depending on availability of electricity. Pumps are always installed at water sources, which are far away from their places. In most of the cases, the way towards pump is hazardous; there are fluctuations in power supply, difficult terrain, fear of animals on the way to pumps, rains, shock hazards etc. Because of these, farmers are avoiding to stop their pumps. This causes loss of valuable electricity & water and ultimately this is loss to the nation. In most of the cases pumps are more than 2 Km to 10 Km away from farmer's houses. In this case they are using vehicles to reach at place of pump to make it operative, which causes additional loss of fuel & time.

CHAPTER 5

Expert System: An ICT Tool for Better Farming

There are many problems in the fields and offices for which we seek the advice of an expert. We recognize persons and experts not only for their knowledge through study but also their ability to apply the knowledge very effectively by virtue of the experience gained over a period of time. The need for experts increases in both dimensions as numbers as well as in the spheres to which expertise is needed. The sphere in which the opinion of an expert can influence decisions or actions is referred to as the domain of the expert. For last few years application of Information and Communication Technology (ICT) has brought revolution in many fields. It is the need of the hour that we explore the possibility and the fields of application in agriculture to acquire maximum benefit from the most promising technology of this information era. ICT is one of the thrust areas in the list given by the planning commission (Now Policy Commission), where emphasis can be given to increase the production and productivity of agriculture commodities. During last 50 years, lot of research has been done in the field of Agriculture, which has not been properly extended to the farmers. ICT could be applied as a means in extending the technologies brought up by National Agricultural Research System (NARS) to the farmers.

Expert Systems (ES) can be used for extending the research to farmers and it can work as a problem-solving tool for them. ES on agriculture could have a powerful mechanism with extensive potential to solve the problems related to agriculture. ES can be developed in any specific domain for some specific purpose. The development of Expert System in different areas of agriculture is an effort in this direction, as it will revamp and vitalize the extension services in order to reach the farmers effectively. The developed expert systems are available on the Internet. So, anyone who is having the internet connection can access the system from any part of the World. For the persons who are not having the internet connection can take the system into Compact Disks (CD's) and can copy it into their computer system. CD's can be obtained from IASRI, New Delhi at a reasonable cost. This expert system will help the farmers in getting instant solution to their problems and it will save time of the experts, they can devote their maximum time to their research work.

Exactly an expert system (ES) is intelligent software which works in one particular domain. It uses knowledge and inference capability to solve the problems. Knowledge residing in knowledge base of expert system consists of domain facts and associated heuristics. This knowledge is collected from human experts who are not available every time. Moreover, one or two experts are not enough for some working areas just like agriculture where so many fields are part of one major field. That is why knowledge is gathered from more than one expert and accumulated to software, which is called expert system. As expert system is based upon knowledge so this is also known as knowledge based system. These systems exhibit the same level of expertise as that of a human expert when used by non-experts. Experiences on the use of ES show that these systems are used as an aid to human beings. ES are not meant to replace human experts or human beings. They only seek to make available the human expertise in a given domain to a wider cross section of users who are not experts in that area.

Daniel Hunt, 1986 concluded that an "Expert System" is an intelligent computer programme that uses knowledge and inference procedures to solve problems that are difficult enough to require human expertise for their solution. The knowledge necessary to perform at such a level plus the inference procedures used can be thought of as a model of the expertise of the best practitioners in the field. Expert system is designed to simulate the problem-solving behavior of a human who is an expert in a domain or discipline. An expert system is normally composed of a knowledge base (information, heuristics, etc.), inference engine (analyzes the knowledge base), and the end user interface (accepting inputs, generating outputs). The path that leads to the development of expert systems is different from that of conventional programming techniques. The concepts for expert system development come from the subject domain of artificial intelligence (AI), and require a departure from conventional computing practices and programming techniques. A conventional program consists of an algorithmic process to reach a specific result. An AI program is made up of a knowledge base and a procedure to infer an answer. One of the most powerful attributes of expert system is the ability to explain reasoning. Since, the system remembers its logical chain of reasoning, a user may ask for an explanation of a recommendation and the system will display the factors it considered in providing a particular recommendation.

Developing an expert system in a specific knowledge domain is quite a difficult task as it requires team of experienced knowledge engineers, programmers as well as domain experts. Agriculture, being a very vast and varied domain of knowledge with over a hundred crops distributed in different geographic regions having varied climatic conditions, building such a team in every domain of knowledge of agriculture is itself a challenging and huge task. Knowledge engineers gather knowledge from domain experts and put it in such a form that system can use for inferring and reasoning using a knowledge representation technique. Programmers then build an online interface so that the end users can use the system over the Internet. To support the Agriculture Extension specialist in developing Expert systems in agriculture IASRI have developed a tool for building online expert system. The tool 'AgriDaksh' was developed at Division of Computer Applications, IASRI to overcome hurdles of expert system development. AgriDaksh is available online through IASRI web site http://www.iasri.res.in or at its direct link http://expert.iasri.res.in/agridaksh/. AgriDaksh enables domain experts to build online expert system in their crops with minimal intervention of knowledge engineers and programmers. With its use, it is possible to build online expert system for each and every crop in significantly less time and resources. Online expert systems have the capability to transfer location specific technology & advice to the farmers efficiently and effectively. This in turn can reduce losses due to diseases and pests infestation, improve productivity with proper variety selection and increase in income of the farmer. Maize AgriDaksh is the first expert system developed using AgriDaksh.

Why Expert Systems?

It's a common consensus that human expertise is scarce so it should be preserve for future uses. Due to scarcity of very best people in the specialized field their replication is also required. Today's organizations have a high-level of know-how experience and expertise which cannot be easily transferred to other members. Expert systems provide very good opportunity to use, transfer and replicate the scarce expertise. The consistency, cost effectiveness and efficiency of the expert system in problem solving open the doors to commercial application of this artificial intelligence in agriculture.

Conceptual Design

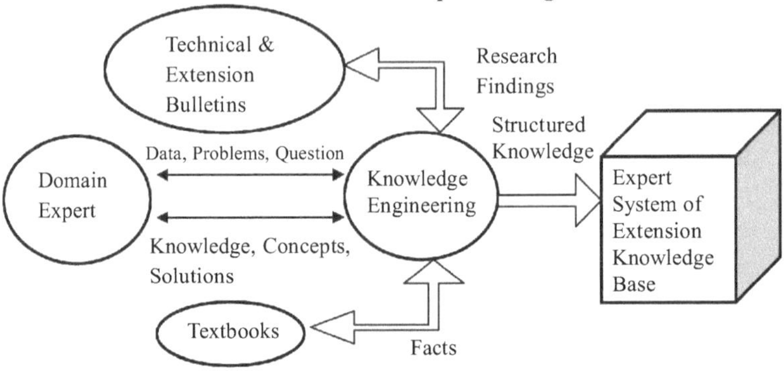

Knowledge Acquisition Module

Some Basic Concepts

Research in AI (Artificial Intelligence) was initially in the direction of developing general programs for the purpose of using them to solve broad range of problems. Very soon, it was found that they could solve only limited number of problems which were similar in nature. This situation led to the development of systems using general methods for knowledge representation, search control and inferencing for a specific domain. It was soon realized that "problem solving power of a program comes from the knowledge it possesses and not from general purpose inferencing". There was a need to provide a program with sufficient and specific knowledge of high quality about some problem area to be able to solve them in that area. This marked the beginning of the ES. Some of the basic concepts involved are defined below:

Artificial Intelligence: Artificial Intelligence is the part of computer science concerned with designing intelligent computer systems, that is, systems that exhibit the characteristics we associate with intelligence in human behaviour.

Expert System: A computer program including expert knowledge to attain high levels of performance in a narrow problem area is called an expert system. Since it possesses large amounts of knowledge and its performance depends on this knowledge, it is also called a knowledge based system. Few examples of expert systems are MYCIN (for medical diagnosis), Prospector (for geological prospecting) and LUNAR (for answering questions about moon rock samples by drawing on a massive database provided by NASA), Rice-Crop Doctor (Expert system to diagnose pests and diseases for rice crop and suggest preventive measures), AGREX (Helps the Agricultural field personnel to give timely and correct advice to the farmers).

Domain Expert: Domain expert is the person who has acquired thorough knowledge of a problem domain through years of training and experience.

Knowledge Engineer: A person who designs expert system and acquires knowledge from domain experts or other sources is called a knowledge engineer.

Knowledge Engineering: Knowledge engineering is the process of building knowledge based systems.

Knowledge Representation: Knowledge representation is the process of formulating or viewing a problem which makes it easier to be coded in a computer language for further reasoning.

Knowledge Base: The component of an expert system which contains domain specific knowledge is called knowledge base (KB).

Inference Engine: Inference engine decides which part of KB (rule) is relevant to the given current situation and then it applies those rules. It is also called control structure. This is domain independent part.

User Interface: This component interacts with user to gather information and gives explanations when required.

How Expert System takes Decision

Knowledge is the most vital part of expert system for decision making. These systems contain large amounts of knowledge to achieve high performance. A suitable Knowledge Representation (KR) scheme is necessary to represent this vast amount of knowledge and to perform inferencing over the Knowledge Base (KB). A knowledge representation scheme means a set of syntactic and semantic conventions to describe various objects. The syntax provides a set of rules for combining symbols and arrangement of symbols to form expressions. Knowledge representation should have:

1. **Expressive Adequacy:** What knowledge can be and cannot be represented in a particular knowledge representation scheme?
2. **Reasoning Efficiency:** How much effort is required to perform inferencing over the KB? There is generally a trade off between expressive adequacy and reasoning efficiency.
3. **Incompleteness:** What can be left unsaid about a domain and how does one perform inferencing over incomplete knowledge?
4. **Real World Knowledge:** How can we deal with attitudes such as beliefs, desires and intentions? Major knowledge representation schemes are based on production rules, frames, semantic nets and logic. Facts and rules can be represented in these knowledge representation schemes. Inference Engines using forward chaining, backward chaining or a

combination there of are used along with these knowledge representation schemes to build actual ES. We will briefly describe these knowledge representation schemes and inferencing engines.

Rule based decision making

A rule based system is also called production rule system. Essentially, it has three parts, working memory, rule memory or production memory and interpreter. Working memory contains facts about the domain. These are in the form of triples of object, attribute and value. These facts are modified during the process of execution. Some new facts may be added as conclusions. Production memory contains IF-THEN rules. IF part contains a set of conditions connected by AND. Each condition can have different other conditions connected by AND or OR. Each condition can give either true or false as its value. THEN part has a set of conclusions or actions. Conclusion may change values of some entity or may create new facts. A rule can be fired when all the conditions in it are true. If any of the conditions is not true or unknown, the rule cannot be fired. If it is unknown, the system will try to determine its value. Once a rule has fired, all its conclusions and actions are executed.

For firing a rule, the system looks into its database. If a rule has some of its conditions satisfied, it is a candidate for further exploration. There may be more than one such rule. This conflict is resolved by some strategy like choosing the rule which contains the maximum number of satisfied conditions, or there may be Meta rules which may be domain dependent to move the reasoning in a particular direction.

Rules may be used in both forward and backward reasoning. When it is used in forward mode, the system starts with a given set of initial data and infers as much information as possible by application of various rules. Again new data are used to infer further. At any point system may ask the user to supply more information, if goal state has not been reached and no more rules can be applied. System keeps on checking for goal state at each firing of rules. Once goal state has been detected reasoning comes to an end.

In backward reasoning mode, reasoning starts with the goal and rules are selected if they have the goal in their right hand side (RHS). To achieve the goal, left hand side (LHS) conditions have to be true. These conditions become new sub-goals. Now the system tries to achieve these sub-goals before trying the main goal. At some point it may not be possible to establish goal by application of rules. In this situation the system asks the user to supply the information.

It may be noted that these rules are not IF-THEN programming constructs available in most of the procedural Programming languages. These are different in the sense that they are not executed sequentially. Their execution depends on the state of the database which determines which are the candidate rules. Another difference is that IF-part is a complex pattern and not just a Boolean expression.

Rules have been used in many classical systems like MYCIN, RI/XCON etc. even today it is the most frequent used knowledge representation scheme. The reason is that most of the time, experts find it easier to give knowledge in the form of rules. Further, rules can be easily used for explanations. One problem with rules is that when they grow very large in number it becomes difficult to maintain them because KB is unstructured. Some techniques like context in MYCIN solve the problem to some extent.

Frame based decision making

The concept of frame is quite simple. When we encounter a new situation, we do not analyze it from scratch. Instead we have a large number of structures or (records) in memory representing our experiences. We try to match the current situation with these structures and then the most appropriate one is chosen. Further details may be added to this chosen structure so that it can exactly describe the situation. A computer representation of this common knowledge is called a frame.

It is convenient to create a knowledge base about situations by breaking it into modular chunks, called frames. Individual frames may be regarded as a record or structure.

Each frame contains slots that identify the type of situations or specify the parameters of a particular situation.

A frame describes a class of objects such as ROOM or BUILDING. It consists of various slots which describe one aspect of the object. A slot may have certain condition which should be met by the filler. A slot may also have default value which is used when the slot value is not available or cannot be obtained by any other way. If-added procedure describes what is to be done if slots get a value. Such information is called facet of slot.

Expert Systems in Agriculture

1. **Rice-Crop Doctor:** The MANAGE has developed an expert system to diagnose pests and diseases for rice crop and suggest preventive or curative measures. The rice crop doctor illustrates the use of expert systems broadly

in the area of agriculture and more specifically in the area of rice production through development of a prototype, taking into consideration a few major pests and diseases and some deficiency problems limiting rice yield. This prototype is results of joint effort by the experts from NIIT and computer professionals of MANAGE while the subject matter expert knowledge on rice pathology and entomology, has been obtained from scientists of Andhra Pradesh Agricultural University (APAU) and Directorate of Rice Research (DRR).

2. **Rapeseed-Mustard Expert System:** R-M Doctor is the Rapeseed-Mustard Plant Protection Expert System Beta version developed using the expert system generic task tools. This system is multilingual and integration of text, image and audio description of symptoms of mustard diseases and insect-pest. Expert system developed to help extension personals, researchers and farmers for identification and management of these diseases and insect-pest. The system is supported by a database containing information about 11 diseases and 10 insect of rapeseed-mustard and 80 color images of various symptoms of diseases and insect-pest. The system involves two main sub-tasks, namely, diagnosis and management. Diagnose sub-task finds out the disease and their cause on the basis of text description and photos of insect-pest and diseases symptoms appears on different part of plant. Management sub-task provides a management plan for the diseases and insect-pest. The R-M Doctor permits only to register user to use the system, guest user can use only sample of identification of diseases of rapeseed-mustard. The aims and objectives of the R-M doctor include: To provide the user friendly online application for identification and management of rapeseed-mustard disease and insect-pest in Indian scenario.

3. **EXOWHEM: Wheat Expert System:** EXOWHEM is a web based Expert System developed for the wheat growing farmers of India. It provides the complete information about the Wheat Crop Management in the country. It advises wheat varieties on the basis of area, cultural and climatic conditions and other characteristics of farmer's interest. It also suggests the appropriate cultural practices like field preparation, fertilizer application, schedule of irrigation etc. It guides them in protecting the crop from insects/diseases/weeds etc. It also provides solution to the problems faced by the farmers through online queries.

4. **IGFRI Expert System:** Indian Grassland and Fodders Research Institute, Jhansi have developed expert systems on fodders and Grasses. The crop on which expert systems are available includes Berseem, Sorghum, Maize, Cowpea, Guinea Grass and Hybrid Napier. The expert systems

provide details information in the form of text and images on each and every aspect of the crops which includes Climate, Soil, Land preparation, Seed bed preparation, Sowing time, Seed cleaning, Seed inoculation, Seed rate, Sowing method, Crop mixture, Fertilizer schedule, Water requirements, cropping system, Cutting management and Forage yield, Weed control, Forage quality, Disease management and Varieties. The expert system can be accessed from website of IGFRI. http://www.igfri.res.in/ fodder_net/html/english.htm

5. **Maize Expert System:** Maize ES was developed by IASRI using AgriDaksh tool. It follows the interaction a user might have with a human expert to solve a problem. It is meant to enhance the efficiency of farmers or Agricultural Extension personnel for maize crop management and to increase the crop yield. It determines the best strategy for irrigating, applying fertilizer and insecticides. Presently, it has four sub-systems: Variety Selection, Cultural Practices, Disease Diagnosis, Insect Identification, and Post Harvest Technology. The Variety Selection sub-system advises location specific varieties and Cultural Practices advises on the aspects of irrigating, application of fertilizers and insecticides. Disease Diagnosis and Insect Identification sub-systems help the stakeholders to diagnos the disease and to identify insects affecting the maize crop and suggest preventive and control measures. Post Harvest Technology sub-system deals with storage and processing of maize for developing value added products.

6. **Expert System on Mushroom:** Expert System for Mushroom Crop has been developed using AgriDaksh. This system is a farmer oriented and user friendly software which provides spectrum of information with images of mushroom crop such as cultivation technology of different mushrooms, diseases and pest management, spawn production technology, compost preparation by short & long method, Post harvest handling & harvesting, Crop management, Nutrition & medicinal value, Transfer of technology, Fungal, viral, bacterial diseases & abiotic disorders etc. Cultivation Technology module gives detail information about compost preparation, spawning, spawns running, casing, fruiting. Post harvest management module gives detail information about Packing and Storage and Short Term Storage. Pest & Diseases module gives detail information about Insect, Pest like Nematodes, mites and springtails and several diseases like Dry Bubble (brown spot), Wet Bubble (White Mould), Cobweb, Green Mould, False truffle (Truffle disease) etc, and timely control measures against pests & diseases.

7. **Expert System of TANU:** Tamil Nadu Agriculture University has developed a number of expert systems includes the crops Paddy, Coconut, Sugarcane, Banana and Ragi and Animals like Cattle and Buffalo, Sheep and Goat, and Poultry. These expert systems were launched by Dr. S. Ayyappan, Secretary, DARE and DG, ICAR on January 25, 2013. Expert systems were developed on a network mode having the network partners of Directorate of Research on Women in Agriculture, Bhubaneshwar, Zonal Project Directorate, Zone VIII, Bengaluru, Tamil Nadu Agricultural University, Coimbatore and Tamil Nadu Veterinary and Animal Sciences University, Chennai. The ES on Agriculture, Horticulture and Animal Husbandry provides systematic and scientific knowledge base on various aspects of the crops, animal production and management.

   Expert System for Rice provide detail expert knowledge on every aspect of paddy production like botany and climate, season and variety, rice eco-system, nursery management, cultivation practices, nutrient management, crop protection, farm implements, post harvest technology, marketing, schemes and institute for paddy , FAQ's on paddy.

   Expert System for Coconut provide detail information on different aspects of coconut like Coconut Planting Seasons, Climate , Varieties, Mother Palm Selection & Nursery Management, Main Field Cultivation Practices, Irrigation Management, Nutrient Management, Pest and Disease management, Coconut Processing, Harvest and Post-Harvest, Farm Implements, Statistics, Marketing, Cost of cultivation, FAQs, DOs and DONTs on the coconut.

   Expert System for Sugarcane provide expert knowledge on every aspect of sugarcane production the sub heading includes Botany & Climate, Season & Varieties, Nursery Management, Cultivation Practices, Sustainable Sugarcane Initiative, Irrigation Management, Nutrient Management, Crop Protection, Farm Implements, Post Harvest Technology, Marketing, Institutions & Schemes, FAQ's on sugarcane production and management.

   Banana Expert System is a comprehensive knowledge base provides scientific Banana production technology to the farmers as well as extension professionals. The various aspects of the technology includes, Botany, Climate, Season & Varieties, Planting Material, Cultivation Practices, Irrigation Management, Nutrient Management, Crop Protection, Farm Implements, Post Harvest Technology, Marketing, Institutions & Schemes, FAQ's.

Expert System for Cattle and Buffalo: This ES is a complete knowledge base which opens avenues for scientific cattle and buffalo production and management. It offer opportunity to the cattle-buffalo owners and dairy extension workers to have technical now how on every aspect of animal production management. It provide expertise knowledge of Breeds of Cattle and Buffalo, Housing Management, Fodder Production, Feeding Management, Breeding Management, Disease and Control, Management Production Technology, Calf Management, General care and management, Common Managerial Practices, Farm Machineries, FAQ's etc.

Expert System on Sheep and Goat: It Make available technologies to the Sheep-Goat owners and extension workers to have technically sound on every aspect of Sheep-Goat production and management. It provide expertise knowledge on Breeds, Housing Management, Fodder Production, Feeding Management, Breeding Management, Young Stock Management, Disease control, General Disease Management, Farm Appliances, Handling and Transport, Vices of Sheep and Goat, Institutions, FAQs etc.

Expert System on Poultry: The Poultry Expert system developed by Tamil Nadu Veterinary and Animal Sciences University provide expert knowledge base on the following aspects of poultry production. General Information about Poultry, House Instruction, Incubation and Hatching, Brooding of Chicks, Grower Management, Layer Management, Disease Control and Management, Breeder Management, Litter Management, Bio-security and Disease Management, Lighting in Poultry Production, Poultry Farm Equipments, Poultry Farm Records Maintenance, Layer Production Indices, Important Vices of Poultry, Broiler Management etc.

8. **Six-Vegetable Expert System (VEGES):** VEGES was developed with the aim of providing diagnosis and treatment for the diseases and disorders of the species most commonly cultivated under cover in the Mediterranean region: pepper, lettuce, cucumber, bean, tomato. Although the VEGES has been designed for use by greenhouse growers of the Mediterranean region, it could equally be adopted by growers in the parts of the world, where similar, simple greenhouses are used for vegetable crops.

9. **Sprayer Nozzle Selection and Sizing Expert System:** This system was designed to assist in the selection and sizing of spray nozzles, covering 275 nozzles and is rated based on five factors. The overall ranking of the nozzles selected is based on nozzle type, operating pressure, nozzle spacing, and boom height and droplet size. A nozzle interchange database is available for five nozzle manufacturers.

10. **Weed Control Systems:** Weed is known as an unwanted plant, characterized by rapid growth. It typically replaces other, more desirable plants. Usually mechanical or cultural methods are used but nowadays along with herbicides or chemicals, different biological methods are also being used to control the weeds and such type of systems give suggestions related to all of these methods.
11. **SEMAGI:** An interactive microcomputer program named SEMAGI has been developed for Sunflower to determine the appropriate selection of herbicides. It combines relational databases on herbicides, weeds and their interaction. The expert system processes and selects the herbicide(s) under the constraints of herbicide efficacy data and of a weed-crop competition model. In addition, SEMAGI provides an economic study of any herbicide treatment selected or introduced by the user, based on herbicide treatment cost, expected yield increase from the weed control treatment and sunflower selling price.
12. **Farm Nutrient Management Expert System:** A computer program was developed to assist farmers in developing customized farm nutrient management plans that reduce crop production costs and environmental risk. The program uses detailed farm information to determine the field by field allocation of manure and to balance this with commercial fertilizer additions.
13. **Farm Planning Systems:** Overall farm planning includes crop storage, protection of farm, cattle management and financial strategies, so all of these activities can be managed at a time in a proper manner by such type of systems.
14. **Expert System to Diagnose Chilling Injury Symptoms on Fresh Fruits and Vegetables:** This system was designed to diagnose chilling injury (exposure to low, nonfreezing temperatures) symptoms, which are not same for all commodities and can be influenced by environmental factors. So we can prevent the market life of certain commodities to be shortened by chilling injury.
15. **Expert System for Soil Erosion Control Planning:** The main purpose of this system was to control soil erosion in agricultural area of Prince Edward Island by recommending appropriate engineering solution. This system was developed at the Technical University of Nova Scotia as a doctoral dissertation using the rule-based PC plus expert system shell by taxes instruments.

16. **Field Master:** The Work Efficiency Institute in Finland has developed this knowledge-based computer program for planning plant production. It helps the farmer to select the appropriate plant and variety for a certain field and to select the appropriate fertilizers and limes. The program guides the user in the direction of sound farming practice by favoring proper crop rotation and by calculating the fertilizer application levels for the different parts of the field according to actual soil properties. This system was programmed with expert object, a rule-and-object based expert system shell running in the PC-environment.

17. **Fertigation Expert System:** This ES is part of CUPTEX (Cucumber Production Management under Plastic Tunnel Expert System), which was developed by Central Lab For Agricultural Expert System (CLAES) within the Ministry of Agriculture and Land Reclamation, Egypt. CUPTEX was developed using a methodology based on KADS and composed of three expert systems: Fertigation, Plant Care, and Disorder Remediation. The main function of fertigation expert system is to give the user a recommendation about the quantity of different fertilizers and the optimum irrigation water quantity that fits his/her situation at any cultivation stage. This model takes number of inputs representing the current situation of the environment including, plant age, soil type, water salinity, climate, and some factors related to the plantation itself such as the drainage quality, the soil sterilization, the sequence of grown cucumber in the same tunnel, the type of fertilizers to be applied, and others factors related to the fertilizers to be used and the expected yield (Rafea, A. 1995). Similarly, soil conservation expert systems have also been made such as.

18. **Disorder Remediation Expert System:** This system is also the part of CUPTEX as mentioned before. The main function of this expert system is to generate a prescription to protect a certain disorder or a set of disorders. In case that the user suspects the cause of disorder(s), he/she can provide the system with his/her suspicion, and the system confirms or rejects this suspicion. If the user has no suspicion, he/she can provide the system with the symptoms of the disorders, and the system identifies the causes of the disorders.

19. **POMME:** This is an expert system for apple orchid management. POMME advises growers about when and what to spray on their apples to avoid infestations. The system also provides advice regarding treatment of winter injuries, drought control and multiple insect problems.

20. **CALEX:** This is a blackboard based integrated expert decision support system for agricultural management, developed at University of California. CALEX can be used by growers, pest control advisors, consultants and other managers. CALEX system has been developed for agriculture management. It is domain independent and can be used with any commodity. CALEX consists of three separate modules: an executive, a scheduler, and an expert system shell.

21. **Agroforestry Expert System:** UNU-AES an agro forestry expert system was designed to support land-use officials, research scientists, farmers, and other individuals interested in maximizing benefits gained from applying agro-forestry management techniques in developing countries. UNU-AES is a first attempt to apply expert systems technology to agro-forestry.

22. **The Central Laboratory for Agricultural Expert Systems (CLAES)** is helping farmers throughout Egypt optimize the use of resources and maximize food production. Some of the Expert Systems developed at CLAES are :

    a) **CUPTEX:** An Expert System for Cucumber Crop Production. CUPTEX currently provides the following services. Disorder Diagnosis, Disorder Treatment, Irrigation Scheduling, Fertilization Scheduling and Plant care.

    b) **CITEX:** An Expert System for Orange Production. CITEX currently provides the following services: Assessment of a farm, Irrigation Scheduling, Fertilization Scheduling, Disorder Diagnosis, and Disorder Treatment.

    c) **TOMATEX:** An Expert System for Tomatoes. The current version of the system (TOMATEX ver. 3), provides recommendations concerning the following agricultural activities: Disorder Diagnosis, Disorder Treatment.

Advantages of Artificial Expertise

1. Human expertise is perishable. If an expert does not practice for a long period, he may lose his expertise. But artificial expertise is relatively permanent and free from human factors.

2. Transferring knowledge from one human to another is a laborious and lengthy process but the same is very easy with artificial experts.

3. Expert Systems are more consistent than human experts. A human expert may miss a rule due to emotional problems in an hour of crisis.

4. Consulting human experts is quite expensive. Also, society spends lot of money to train them. Using expert systems is much cheaper.
5. Development cost of an expert system is quite high but when it is replicated, the cost comes down. Similarly, operational cost is also quite low.
6. Provide expert level recommendations understandable to users
7. Have an ability to imitate human through and reasoning.
8. Maximizes benefit of the agriculture farm.
9. Facilitate efficient use of available resources and infrastructure.
10. Makes aware about cost benefit ratio before actual adoption.
11. An efficient toll for appropriate decision making/taking.
12. Encourages farmers for diversification and quality production.

Limitations of Expert Systems

1. They are not creative like human experts. Thus expert systems do not use novel approaches. Human can learn new problem solving knowledge and adapt to new situations but expert systems cannot handle unexpected situations.
2. Expert Systems do not have sensory input mechanisms. All the information has to be converted into symbolic form and then only can it be given to expert systems. In some situations this may be a severe limitation such as a case where the saying 'a picture is worth a thousand words' applies.
3. Expert Systems focus on a narrow domain. They cannot handle many issues which are not very likely to occur but may affect the situation. To incorporate knowledge about these issues may mean solving thousands of problems together of an expert system?
4. ES is expensive because it requires some investment in PC with internet connectivity.
5. Time consuming.
6. Required trained experts.
7. Problems in communication of superior expert system.

Conclusion

As an economical aspect, expert systems can clearly affect over all agricultural production. Similarly, we can preserve our natural resources instead of being polluted Because of excessive and wrong usage of chemicals and fertilizers. Moreover, better training of extension workers through these systems can also be beneficial. These systems are completely computer based but the people living in third world counties are not fully computer literate. Hence the use of expert systems in these countries is still a dream. That is why there is also a need to work in this regard. The expert system is very good tool for farmers & extension workers but the ES must be developed in local languages so that it could be helpful to the farmers to develop their own expertise to enhance the production and productivity of Agriculture. These ES must be available in the village booths so that it may be used by the needy persons.

References

Expert Systems – Introduction to First and Second Generation and Hybrid Knowledge Based Systems, Chris NIKOLOPOULOS *Bradley university peoria, Illinois.* ***Publisher:*** *CRC (January 10, 1997)*

Practical Applications of Expert Systems, Susan Lindsay *182 pages* ***Publisher:*** *John Wiley & Sons Inc (September 1988)*

The Organization of Expert *Systems, a* Tutorial Mark Stefik, Jan *Aikins,* Robert Balzer, John Benoit, Lawrence Birnbaum, Frederick Hayes-Roth, Earl Sacerdoti Can be found at URL.

http://www.manage.gov.nic.in

http://en.wikipedia.org/wiki/Expert_system

http://en.wikipedia.org/wiki/Mycin

http://www.cee.hw.ac.uk/~alison/ai3notes/section2_5_5.html

http://www.epa.gov/oppt/newchems/tools/oncologic.htm

http://www.epa.gov/oppt/newchems/pubs/sustainable/onco-faq.pdf

http://www.generation5.org/content/2005/Expert_System.asp

http://www.claes.com

http://www.passaoc.org

CHAPTER 6

Knowledge Management

"Knowledge is Power" – ***Francis Bacon***

"Knowledge is proud that he knows.
Wisdom is humble that he knows no more"- ***Anonymous***

Introduction

The world is said to be a global village organizations and people must learn, unlearn and relearn to expand their capacities and survive in the highly competitive market place. In view of rapidly changing socio–economic circumstance, it is a challenge to survive, excel and grow. It is knowledge which can make difference in getting the competitive edge. Thus, organizations need to pool their internal knowledge and find ways to update, enhance, share and create value. Organizations need to analyze their collective knowledge and use them most efficiently. Need for new and new technology is driving people to innovate in every sphere. Thus, management of knowledge has become the buzz word today. *Knowledge Management deals with process of creating, updating, sharing, storage and dissemination of knowledge within the organization*. Use of information and communication technologies is common to speed up the process. It is the need of hour to have knowledge repository in every organization for continuous learning and growth.

In the current era of knowledge-driven society, knowledge is a critical resource, which needs to be leveraged and utilized optimally. Information and communication technology (ICT) has been the key in the knowledge management process. A technology-driven organization needs to leverage knowledge management process efficiently to be effective and competitive. Such organizations tend to capture implicit (tacit) knowledge using various IT-based technologies like data mining and make this knowledge explicit and available in public domain for intellectual consumption. Information repositories like knowledge centers or libraries have begun to appreciate the power of the knowledge management systems and are investing their critical resources in terms of manpower and technologies to make them technology driven. The functions of information professionals are undergoing drastic changes in terms of moving beyond cataloguing and sorting to include providing customized information to individual requirements. Therefore, information professionals, today, need to be intellectually equipped to leverage knowledge management systems to provide required customized services efficiently.

What is Knowledge Management (KM)?

Knowledge is information in our mind and tested in practice. Knowledge management (KM) is a very rapidly developing area in which the information professionals have a critical role. Significant changes are required in their role, skills, education and training so that they can confidently face the challenges of KM in the changing digital environment. Gartner Group defined knowledge management as a "discipline that promotes an integrated approach to identifying, managing and sharing all of an enterprise's information assets. These information assets may include databases, documents, policies and procedures as well as previously unarticulated expertise and experiences resident in individual workers. knowledge management issues include developing, implementing and maintaining the appropriate technical and organizational infrastructures to enable knowledge sharing" *According to Liebowitz and Beckman: "knowledge management covers identifying what knowledge assets an organization possesses, analyzing how the knowledge can add value, specifying what actions are necessary to achieve better usability and added value, and reviewing the use of the knowledge to ensure added value."* Thus, knowledge management is a strategy to provide the right knowledge to the right people at the right time and to help people share and utilize this knowledge so that organizational performance can be improved. KM facilitates creation, access and reuse of the knowledge. It involves, finding out what an organization knows; where the knowledge resides in the organization; how to locate people with specific knowledge, expertise and experience; and how to share and utilize the accumulated knowledge. Various practitioners and research scholars have defined knowledge management in different ways but the main emphasis remains on the sharing of information assets of organizational knowledge. Knowledge management has become crucial tool for managers. It is needed to create value and enhance effectiveness of people, systems and organizations. Knowledge management caters to the critical issues of organizational environment. It embodies organizational process to seek synergistic effect by combining information technology, creative and innovative capacity of human beings. The goal is to sustain through learning and adaptation.

Data, Information and Knowledge

Data versus Information: Data refers to raw, unevaluated facts, figures, symbols, objects, events, etc. Data may be a collection of facts lying in storage, like a telephone directory or census records. Information is data that have been put into a meaningful and useful context and communicated to a recipient who uses it to make decisions. Information involves the communication and reception of intelligence or knowledge. An element of data may constitute information in a specific context; for example, when you want to contact your friend, his telephone number is a piece of information; otherwise, it is just one element of data in the telephone directory.

Information: This is an era of information's power. All information is neither relevant nor desired. Right information at the right time to right people becomes meaningful and precious. The term information refers to facts or data which have been organized for a specific target audience. It consists of a pattern imposed on data which affects the interpretation of these data and enables them to be transmitted Example: Weather data of days, weeks & months in terms of parameters of weather can be helpful to decide agricultural operations. The chronologically listed marks obtained by students in a course in different tests over a semester refer to the progressive increase or decrease in marks as well as overall performance of students in the course.

Characteristics of Information: The characteristics of good information are relevance, timeliness, accuracy, cost-effectiveness, reliability, usability, exhaustiveness, and aggregation level. Information is relevant if it leads to improved decision making. It might also be relevant if it reaffirms a previous decision. If it does not have anything to do with your problem, it is irrelevant. For example, information about the weather conditions in Jaipur in January is relevant if you are considering a visit to Jaipur in January. Otherwise, the information is not relevant.

Timeliness refers to the currency of the information presented to the users. Currency of data or information is the time gap between the occurrences of an event in the field until its presentation to the user (decision maker). When this amount of time is very short, we describe the information system as a real-time system.

Accuracy is measured by comparing the data to *actual* events. The importance of accurate data varies with the type of decisions that need to be made. Payroll information must be exact. Approximations simply will not suffice. However, a general estimate of how much staff time was devoted to a particular activity may be all that is needed.

Knowledge: Knowledge is primary cognitive content of a culture. When people interact with environment, they gather experience. How this experience is gathered, documented, verified and applied differs from culture to culture. Knowledge in case of agriculture refers to crops, soils, livestock, climate, cultural practices and management. Knowledge may be held in a book, report, film or in minds of the people. Knowledge can be put in many categories such as concepts, theories, artifacts, etc.

According to Rolling: Knowledge is a property of the mind and cannot be communicated to others unless transformed or encoded knowledge proud is intrapersonal. Knowledge is organized body of information for comprehension and understanding. Knowledge includes concepts, theories and objects.

Explicit and Implicit Knowledge

Knowledge is of two types such as explicit knowledge and implicit knowledge. Explicit knowledge (knowledge that can be codified) is about facts available in books or records this kind of knowledge deals with 'what', examples of explicit knowledge are databases, data warehouses, data marts, documents, software codes, etc. to the other end implicit knowledge (knowledge that remains in one's mind and cannot be codified) is intuitive dealing with 'how', examples of implicit knowledge include techniques and insights gained from personal experiences and interactions. It is easy to find explicit knowledge but very difficult to get implicit knowledge. Tacit knowledge is gained through experience, analyses and reflection and cannot be shared as such, implicit knowledge is action–based and rooted in practice and therefore, it cannot be easily explained or described. Tacit knowledge is something that cannot be converted into explicit knowledge. We know more than we can tell. We cannot articulate all that we know because we are not fully conscious of all the experience. It resides in human mind.

Wisdom: Wisdom is specialized & integrated knowledge. It is the ability to use one's knowledge and experience for making good decisions and judgments. It is enlightened understanding and very much individualized application of knowledge to produce results.

Why Knowledge Management

In this knowledge era, our success depends on what we know and what we do with what we know. The changes that we are going through are overwhelming. Following are the primary reasons why knowledge organizations should implement Knowledge Management (KM) strategy and manage their collective intellectual capital to succeed in this challenging knowledge era.

- Managing the complexities of knowledge, knowledge exchange and knowledge development to addressing the pain points or challenges of 21st century organizations.
- Accelerating learning, innovation and execution to maximize top line and bottom line.
- Managing the continuous changing of business needs and the way business is managed.
- Placing the true potential on a springboard.
- Increasing productivity and profitability while organizations are going through margin pressure due to tuff competitions.
- Addressing the highest expectations of customers.
- Meeting the market demands in this continuously changing business environment.
- Increasing need for building and accessing common knowledge easily and quickly.

Why Knowledge Management in Agriculture

Knowledge plays a significant role whenever change, innovation and growth are being pursued in a competitive and complex field. Agriculture today is just such a field. Leveraging knowledge is thus a critical input in the transformation of agriculture from subsistence to market-oriented economic sector. A demand-driven agricultural knowledge management system facilitates access to adoption of appropriate technologies and processes from research and development institutions. The resulting benefits are many including:

- Being adaptive to changes in the Agri-food chain.
- Learning from organizations with good research output and innovations record.
- Benefiting from indigenous knowledge.
- Being responsive to farmers and pastoralists needs and requirements.
- Addressing inertia due to lack of information among farmers and pastoralists.
- Creating a stable technology infrastructure over a long period of time.

- Creating a capable agricultural management regime with a good development process supporting them, and
- Developing a loyal and capable workforce.

If we agree on the premise that development of a demand-driven knowledge management system will have a positive impact on the development of agriculture, the next logical question will be how we develop such as system.

Steps Involved in Knowledge Management Process

Knowledge management processes are the activities put in place to enable and facilitate the creation, sharing and use of knowledge for the benefit of the organization. The main processes are knowledge generation, knowledge storage, and knowledge utilization. These are carried out in a spiral way. After knowledge is created/acquired, stored, shared and utilized, it goes to next phase of enhancing / refining of existing knowledge, and identifying and acquiring new knowledge. Though knowledge management processes involve many steps, knowledge creation and sharing are the primary concerns of a knowledge management programme.

Studying knowledge processes is valuable in order to understand the roles and functions of the different actors involved at various phases in which knowledge is transformed from its abstract to the more concrete and applicable forms. Generally, there is no as such demarcation among these steps in the process. For example, in farmers' experimentation knowledge can be generated and exchanged without undergoing any transformation. Moreover, for these processes to take place, building and strengthening supporting institutions are necessary and the understanding of the production, reproduction, and transformation of knowledge must be situated in the individuals and groups involved in the process. The processes involve certain steps as detailed below:

a. Knowledge Generation/Production/Creation

It comprises knowledge identification, acquisition, capturing, and synthesis. Knowledge identification is essential to know what knowledge the organization and its people require in order to meet their goals and objectives. This can be done by surveys, interviews, group discussions and brainstorming. Knowledge acquisition involves bringing knowledge into the organization from external sources. It deals with acquiring knowledge skills, theories and experience needed to create core competencies and knowledge domains. Use of ICT can expedite knowledge acquisition at a low cost. Knowledge capturing requires capturing by implicit knowledge of the organization gained and built through years of experience. This knowledge has to be captured by proper documentation, through mentoring, training and surveys. Knowledge synthesis is analysis of existing knowledge and discussion with various experts on a specific subject or issue. The knowledge synthesized is more accurate, complete and enhanced. Synthesis of knowledge can be a useful tool for the preparation of new research projects, solving various problems, and for creating a more comprehensive, overall perspective on a specific subject.

Who generates knowledge? Knowledge can be generated by multiple actors. Farmers, researchers, change agents, policy- makers, project managers, investors, system scientists; national and intentional NGOs, women's groups etc. This process is typically attributed to research.

b. Knowledge Storage and Retrieval

These processes seem to be the task of specialized libraries. But scientists, extension workers & farmers also store & retrieve information in forms of books, magazines, compact Disc, videos, photographs and other modern ICT tools. The acquired knowledge should be codified and stored in the databases and knowledge warehouses, where it can be easily accessed and utilized by the organization. Knowledge needs to be organized for storage and retrieval. It entails knowledge typology, knowledge bases and knowledge maps. Knowledge typology can be classified into implicit and explicit knowledge. Tacit knowledge is experimental and localized in people's mind, where as explicit knowledge is available in documented forms. Knowledge base is the transformation of databases to knowledge bases through information bases. It stores both structured and unstructured contents like electronic documents. Knowledge base facilitates the storage and sharing of explicit knowledge. Knowledge map is a directory that relates people who need knowledge to the places where it can be found. It discovers the knowledge resources within the organization and guides people by providing location information for the most relevant knowledge resources. This involves the discovery of implicit knowledge for sharing.

c. Knowledge Transformation (Oral, Written, Simple, Meaningful, Relevant).

Transformation: The essence of Agriculture Knowledge Information System (AKIS) is that knowledge generated in one part of the system is turned in to information for use in another part of the system. The transformation process taking place in AKIS are as follows:

- From information on local farming systems to research problems.
- From research problems to research findings.

- From research findings to tentative solutions to problems.
- From technologies to prototype recommendations for testing in 'farmers' fields.
- From recommendations to observation of farmer's behavior.
- From technical recommendations to information affecting service (inputs & marketing) behavior.
- From adapted recommendations to information dissemination by extension.
- From extension information to farmer knowledge.

d. Knowledge Exchange

Exchange of information between the members of the clientele groups can happen in a number of ways depending on the social structure and relations. In villages people meet in market places, field and homes or at bathing place and may exchange information about the problem areas. They may even look at the field and ask about the techniques. The ways and means of exchange are important from point of view of extension. It is common knowledge that messages are shared mostly through farmer to farmer communication.

e. Knowledge Dissemination

Knowledge dissemination is communication of knowledge, concepts from the place of origin to the intended users. Dissemination may take place through various means such as word of mouth or written communication or other means of information and communication technology (ICT). This is a process in which various actors and agencies are involved. Even though extension has the responsibility to communicate development messages, there are other agencies simultaneously involved. Farmers prefer to communicate with farmers.

f. Knowledge Utilization

This is stage in which knowledge affects thinking and action of the receiver. The receiver may use the knowledge to confirm or reject. The message may be used on a short plot in the beginning and can be continued. Knowledge utilization can be achieved through knowledge dissemination, sharing and application. Knowledge dissemination is the process of dissemination of knowledge to others. Intranet, knowledge bases and expert databases can be used to facilitate dissemination of explicit knowledge. This can be through point-to-point dissemination, knowledge broadcast or through enabling search to knowledge bases. Knowledge sharing means retrieving knowledge from the knowledge repositories and making it available and accessible to users. Tacit knowledge is shared through interaction among people working together in an organization while explicit knowledge can be shared through databases, expert systems, knowledge bases, knowledge warehouses, etc. Knowledge can be shared with all on need basis using the above means. Knowledge application involves using the retrieved knowledge for solving problems, performing tasks, making decisions, researching ideas and learning. There are many technologies, which are integrated with the aspects of acquiring, generating, storing, organizing, sharing and disseminating knowledge that capture, acquire Knowledge & affects individual and organizational performance. Artificial intelligence, expert systems, information retrieval technology, electronic publishing technology, computer-supported collaborative work, groupware, decision support systems, database technologies, help-desk technology, brainstorming applications, performance-support systems, simulation softwares, document management, web mapping tools are few technologies which are connected to knowledge management.

The Three Pillars of Knowledge Management

a. **People (Knowledge Workers):** Knowledge worker is an individual with the ability to think, learn and perform. The quality of knowledge worker is measured in terms of innovation, excellence, speed and quality. They are responsible for knowledge creation and application. Individual capability, energy level and interest are addressed here.

b. **Process (Best Practices):** It drives collaboration or teamwork and brings consistency and scalability to some extent. Collaborative culture and system thinking are addressed through knowledge management practices. An innovative approaches for performing certain task or function in the best possible way.

c. **Product (Platform):** A technology or a systematic approach that can enable knowledge management. Sophisticated or innovative system must be process-driven and must be easy to embrace. Comprehensive system maximizes the scalability and consistency and help people embrace the knowledge management practice naturally. Comprehensive knowledge management systems in line with the business functions or business objectives must be addressed here.

Role of Information Professionals in Knowledge Management

The conventional role of information professionals was to collect, process, disseminate, store and utilize information to provide multi-disciplinary services to the personal and professional needs of the users. They were also involved in a continuous search for excellence and codifying information sources. But now their role is not restricted to information management only. They play major role in knowledge management programmes and in identifying, acquiring, developing, resolving, storing and sharing of knowledge. In addition to satisfy the information needs of the users, they have to understand the knowledge needs of the organization and its people also. They have to create an approach to transform and share implicit and explicit knowledge and enable utilization of knowledge. The information professional now needs to develop knowledge about the nature of the knowledge required by the organization and knowledge behaviour of the people and how people acquire knowledge, i.e., how people share, disseminate and utilize knowledge. They should know about the needs, role and value of the knowledge in the organization. Kim stated that "Knowledge managers (information professionals) should shape the knowledge policies, structures, processes and systems that will nurture organizational learning. They should be able to extract, filter and disseminate vital external knowledge". Shanhong opined that "knowledge management in libraries should be focused on effective research and development of knowledge, creation of knowledge bases, exchange and sharing of knowledge between users, training of staff, speeding up explicit processing of the implicit knowledge and realizing need of its sharing".

Information professionals have to manage relationships with external providers of information and knowledge and should negotiate with them. They should design, implement, and oversee knowledge infrastructure, computer knowledge networks, and content management of the organization's Intranet. They have to provide critical input to the process of knowledge creation and its use and should continuously evaluate its relevance to the organization. They have to design information systems, which will ensure ongoing knowledge flow from external and internal sources. They have to identify barriers and priorities changes, which will enhance the spontaneous flow of knowledge throughout the organization. They should work side by side with users in collecting and analyzing strategic intelligence, and should act as trainers and consultants who transfer knowledge gathering and research skills throughout the organization.

A bell and Oxbro stated that "Information professionals have to identify, acquire and evaluate internal and external sources of knowledge and integrate, organize and make relevant knowledge available to the right person at the right time". Educating people the advantage of sharing knowledge, experience and expertise is an important task that information professionals have to perform. Staff should be trained to use a wide range of retrieval and analysis tools. Guns have laid down following role and responsibilities for the information professionals (knowledge managers) in knowledge management.

- Warehouse and track information and maintain a knowledge inventory.
- Ensure the right kind of technology is in place such as Internet, Intranet, databases, and library softwares, OPAC, etc.
- Identify valued skills, knowledge and expertise in the organization.

Carrillo proposed three main roles for the knowledge professionals: Managing knowledge repositories; facilitating knowledge flow and communication; and leveraging value generation capacity. All KM strategies are likely to embrace elements of all three of these approaches. Not one of these can be achieved without some attention to the others. Overall, informational professionals have to perform the following roles to leverage knowledge management programmes:

- Management of contents from licensed online resources like e-books, e-journals, databases, etc.
- Analysis of documents, organization, classification and sorting, and their archiving for easy retrieval and faster dissemination.
- Creating new knowledge through filtering, consolidating, and repackaging information.
- Understanding users need, helping them to locate and retrieve information of their own.
- Training of users to maximize use of the knowledge repository.
- Communication in the hierarchical structure of management to keep track of the process of workflow in the organization.
- Digitization's of various in-house documents like project and research reports by students and faculty.
- Development/deployment of tools for information customization.
- Imparting information literacy instruction to the users.

Skills Requirement of Information Professionals

Information professionals should possess excellent communication skills, and abilities to offer specialist skills such as expertise in electronic systems and resources or experience of planning and delivering training. They should have skills in persuasion and reasoned argument in a changing and complex organizational environment. The various skills needed by information professionals in knowledge management programme are:

- Interpersonal communication skills for transferring implicit knowledge to explicit knowledge. It also encourages people to identify and share relevant ideas, knowledge and information.
- General management skills such as leadership, human resources management, change management, project management, etc. are seen as a vital part of knowledge management programme that information professionals should have.
- Information management skills like classification and cataloguing, indexing and abstracting, information consolidation and repackaging, content management, storage and retrieval, construction of controlled vocabulary and thesaurus.
- Information technology skills like webpage development, database design and management, web publishing, Internet publishing, designing and application of search engines, networking to gather, share and disseminate knowledge.
- Other skills such as, strategic thinking, writing, teaching and learning skills, presentation skills, skills to convert information into knowledge, and skills to understand knowledge management principles and processes are also very much essential for knowledge management programme.

Challenges for Information Professionals

Following are challenges for the information professionals.

- Management of implicit knowledge is not so easy. It is hard to know what is in human mind. It is also difficult to capture knowledge and manage it within large organizations.
- People hesitate to share their knowledge because of insecurity and fear of passing their implicit knowledge to colleagues and lose their importance. Overcoming this fear and motivating such people is the biggest challenge of knowledge management.
- The ever changing pattern of information generates huge amount of information in short span of time. Thus, information gets obsolete as soon as it is generated. Because of this, it is difficult to provide nascent information quickly to the user.
- If obsolete information is not removed it leads to information overload, which creates problem in identification of useful information.

- Rapid change in ICT handicaps the librarians to communicate knowledge to the users. Since ICT is an important tool for knowledge management, information professionals need a high level of technical skills in ICT.
- Financial constraints are there for knowledge sharing incentives. knowledge management should show overall economy.
- There are so many knowledge management needs; it is hard to know where to start.
- Lack of up-to-date knowledge as to what is happening in ICT and knowledge management.

Benefits of Knowledge Management

Knowledge Management helps develop competitive advantage and maximize productivity, quality and profitability by facilitating collective thinking, innovation, learning and execution.

Benefits to Organisations

- Increases the collective capability of organizations to maintain leadership position.
- Improves operational performance and maximizes organizational ability.
- Accelerates learning and innovation through sophisticated knowledge management processes.
- Facilitates knowledge convergence and enrich the corporate memory.
- Minimizes reinvention and knowledge evaporation.
- Drives best practices for "creative and productive working environment"

Benefits for Enterprises

- Maximizes brand-image and status in the market place and helps attain and sustain leadership position.
- Helps understand customers better; deliver quality and timely response to meet the highest expectations.
- Helps develop and retain employees & partners and maximizes their collective mind power.
- Helps create a connected organization by linking business functions and related knowledge to maximize productivity and profitability.

- Empowers development, sales, marketing, support and other business functions to make it more effective and valuable.
- Helps educate customers on the continuously changing environment and make them self-sufficient.

Benefits for Educational Institutions

- Helps transform the educational institutions into connected "knowledge centers" and expands the campus around the globe.
- Helps manage distance learning and e-learning effectively.
- Brings together teachers, students and alumni across campuses and enables collaboration for collective learning.
- Helps simulate experiential learning model to gain more insights into the subject area.
- Improves students' self-sufficiency, confidence level and learning potential by improving the quality of learning programs to help face the real-life challenges.
- Empowers faculties to deliver high quality learning programs in a smart and scalable way.

Benefits for Governments

- Helps serve citizens and make them smart to maximize nation's wealth.
- Improves national wealth and standard of living through intellectual capital.
- Reduces frustrating information searches and the cost of various functions.
- Connects departments and various ministries to increase their efficiency.
- Helps manage knowledge community initiatives of e-governance to transform into "knowledge society".
- Helps develop "Knowledge Centers" for agriculture, education and administration.
- Selecting appropriate knowledge management tools that serve as enablers to implement the bigger concept.
- Understanding and adopting or responding to the role of organizational culture to the success of knowledge management.
- Appreciation of the value of starting with a pilot implementation.

Conclusion

In the recent past knowledge management has emerged as a powerful strategy for enhancing the productivity as well as creating a sustainable competitive advantage for the organization in the global market. Information professionals can play an important role while managing the explicit knowledge. To achieve this, they need to be equipped with certain additional skills along with the conventional skills. The new role of information professionals is full of challenges and they need to strive hard to be relevant in anticipation of resurgence of the new areas of knowledge management.

References

Abell, Angela and Oxbro, Nigel, 2001. Competing with knowledge: The information professionals in the knowledge management age. Library Association Publishing, London.

Carrillo, F.J. 2002. Capital systems: Implications for a global knowledge agenda. *Journal of Knowledge Management,* 6(4): 379-99.

Davenport, Thomas H. & Prusak, Laurence, 1998. Working knowledge: How organizations manage what they know. Harvard Business School Press, Boston, 1998, pp. 114-15.

Dutta, D. 2000. Managing knowledge within the firm: The enabling role of information & communications technologies. *In* Second International Conference on Business & Management, 2000, Nirma Institute of Management.

Education in and for new times. *University News,* 2004, 42(26):16.

Gartner Group. Building knowledge powerhouse. *Dataquest*, 1998. Liebowitz, Jay & Beckman, Tom. Knowledge organisations. St. Luice Press, Boca Raton, 57 p.

Guns, Bob, 1998. The chief knowledge officer's role: Challenges and competencies. *Journal of Knowledge Management*, 1(4):316.

Kim, Seonghee, 1999. The roles of knowledge professionals for knowledge management. *In* 65th IFLA Council and General Conference, 20-28 August 1999, Bangkok, Thailand. (Available at http://www.ifla.org/IV/ifla65/ papers/042-115e.htm)

Rowley, Jennifer, 2003. Knowledge management– the new librarianship: From custodians of history to gatekeepers to the future. *Library Management*, 24(8/9).433-40.

Shanhong, Tang, 2000. Knowledge management in libraries in the 21st century. *In* 66th IFLA Council and General Conference, 13-18 August 2000, Jerusalem, Israel. (Available at http://www.ifla.org/IV/ifla66/ papers/057-110e.htm)

Streatfield , D. and Wilson, T. Deconstructing Knowledge Management. *Aslib Proceedings*, 1999, 51(3): 70.

CHAPTER 7

Websites and Web Portals: Dedicated ICT Tool for Agriculture Extension

The vision 2020 document of the Department of Agriculture and Co-operation envisages that "the tools of ICT will provide networking of Agriculture Sector not only in the country but also globally and the Centre and State Government Departments will have reservoir of databases" and also "bring farmers, researchers, scientists and administrators together by establishing "Agriculture Online" through exchange of ideas and information". To get realize the vision 2020 a large number of agriculture based websites and web portals have been established in India which is working for wellbeing of farming community. Majority of agricultural websites and web portals are providing need based technological information to the farmers as well as extension professionals. To revitalize our extension system these web based services can be used as ICT tool by extension personnel and farmers also. Therefore, to get access services from these websites and web portals or to replicate these ideologies in other part of country, elementary knowledge of working process and services available on these websites & web portals is essential. This chapter will provide a brief description about major ICT based projects dedicated to Indian farmers.

Website Versus Web Portal

A Web portal is also a type of website but it differs in content and services from a typical website that provides only specialized information. A web portal is a launch pad to a host of web based services such as email, shopping, gaming, news, weather and so on whereas a website is concerned with providing information about a company only. Web portal is indeed a type of website, but it is much more in content and services. A portal is defined in the dictionary as a gateway or an entry point to a grand entrance. Thus a Web portal, in addition to being a website, also acts as a gateway to the internet. For example www.google.com is a website but it also acts as a portal as it gets to you an array of web services. There are many pages that go into making it a website that is also a Web portal as you can use it as a launch pad to millions of other web pages, go shopping, play games, read your email, or talk to your friends through its messenger called Google talk and so on. A Web portal is not just a great source of information; it allows data retrieval from divergent sources. While a Web portal find information for the users, they themselves have to search for it on a particular website. So whereas Yahoo.com or Google.com is considered as web portals, CNBC.com or CNN.com or BBC.com is considered as websites. You can search for information on a website but it is limited while a Web portal itself searches and presents all the information for the users.

SMS Portal for Indian Farmers

ICT is an important tool to reach out to the farmers in a timely, holistic and extensive manner. There are large number of web-based Portals and applications but, due to severely limited penetration of internet in the villages, most farmers cannot benefit from web access as the outreach through Government Offices, Common Service Centres and Internet Kiosks is also limited as far as web enabled services are concerned. However, mobile telephone density in the rural areas is increasing everyday. More than 35 crore mobile connections being used and therefore, the National e-Governance Plan (NeGP-A) and 12th Plan Document for the National Mission on Agricultural Extension & Technology lay great emphasis to provide extension services through mobile phones which gives a potential outreach to nearly 13 crore farm families. During celebrations 85th ICAR Foundation Day, Hon'ble President of India launched on 16.07.2013 a SMS Portal for Farmers created by the Department of Agriculture & Cooperation, Government of India. Officers, Scientists and Experts of Department of Agriculture & Cooperation, Indian Council of Agriculture Research, Department of Animal Husbandry, Dairying & Fisheries, India Meteorological Department and Food grain Procurement Agencies will use this portal for disseminating relevant information, giving topical & seasonal advisories and providing services through SMSs to farmers in language of the State. The State Governments and their field formations down to the Block level (including State Agriculture Universities and KVKs) also have come forward in a big way to use this integrated Portal. Considering availability of more than 35 crore mobile connections in the rural area, Kisaan SMS Portal is likely to be very useful to farmers and all other stake-holders as timely and relevant information will be provided to farmers in their own language.

SMS, USSD, IVRS, Voice Recognition etc. are technological tools which can be successfully implemented through mobile telephony. SMS Portal for Farmers is first such step in this direction and it aims to have an integrated portal for the entire nation for Departments, Organizations of the Central Government and State Governments. Currently officers and scientists of Agriculture, Horticulture, Animal Husbandry & Fisheries Dairying sectors in all three departments under the Ministry of Agriculture (DAC, DARE and DAHDF), ICAR, Indian Meteorological Department, and Agriculture Universities & Storage Organizations have been included. The steps involved for using this web based interface are described in the following paragraphs.

How Does SMS Portal Work?

Initial phase consists of creating a farmer-database and grouping them according to their geographic locations. Farmers can register for the service by filling a physical form with the State/ District Agriculture departments or through internet/ mobile interfaces. In addition to collecting basic information about geographical location and land holding etc., farmers also specify up to eight crops/ agriculture related activities for obtaining information or advice. They are also asked about their preference on receiving these as SMS or Voice messages and their language preference. The system supports sending SMS in regional languages. In case the farmer's mobile does not support regional languages he can give option to receive the SMS in regional languages written in roman script. Officials and scientists who are registered as users can send SMS through various options by selecting mobile numbers from the database, or from an excel sheet or they can add one or more mobile numbers manually. They also have option to select a part of his area down to the block level. Thus, a State level officer can send SMS to the entire State or one or more zones/ divisions districts or some of the blocks in the State. Content of SMSs and number of farmers benefitted are saved for any further evaluation or review at a later date. The SMSs are also sent automatically to other users of the same State, so that they can also view the same for applicability in their jurisdiction. In order to have a holistic view of the system, a comprehensive dashboard has also been developed. It is possible to drill down to the block level and analyze as to what all SMSs have been sent by different departments and how many farmers have been benefitted so far.

i. In order not to disturb the farmers during night, SMS Gateway remains in-operational from 9.30 PM to 7.00 AM every day.

ii. For Hindi speaking States, simple and spoken Hindi should be used.

iii. Unnecessarily long messages should not be sent as every 160 characters in English and about 70 characters in Indian languages is equal to one message, each of which costs money.

iv. Broadcast messages like market prices should be made crop specific as per 8 priorities of the farmer till the time a Pull SMS service is introduced for this purpose.

v. Search facility must be used to see previous advisories to avoid duplication and contradiction.

Registration of the User: The officers who are designated to use this portal can register themselves at the website http://farmer.gov.in. In order to ensure that only genuine users can register, a three tier security level (e-mail, SMS and activation) has been implemented. A simple form to fill in the details is required to fill up. Special care needs to be taken while filling in the email or mobile number as subsequent authentication is linked to the same. After completion of registration process from the user side and user profile is to be activated by the admin user concerned now. As there are various organizations using the system, a hierarchical structure for activating the users has been implemented. While DAC will activate admin users of each Organization/ Department, the admin users concerned will, in turn, activate their users.

Deregistration as user: When users hand over the charge to others; he has to deregister from the portal. Such deregistration request would also be pending for approval and viewable below the pending activation requests. These can be examined by the admin user and deregistration can be completed.

Registration of Farmers: Farmers can register through Kisan Call Centers (1800-180-1551), web portal www.kisan.gov.in or by sending SMS at 51969 or 9212357123. Procedure and format for registration by SMS are as under:

Format for typing in the message "KISAAN GOV REG <NAME>, <STATE>, <NAME>, <DISTRICT>, <NAME>, <BLOCK>, >NAME>" (only first 3 letters of State, District & Block Names are needed). In case of any problem the farmer can send an SMS for help as "KISAAN GOV HELP". He/she will be called back by the Kisan Call centre to help in registration. A text message with all details will be sent back to the farmer. In addition, existing data-bases available with government departments, organizations can also be merged with the database of the Kisan SMS.

Stop SMS Service by Farmers: A registered farmer can stop his SMS service by simply sending a SMS on 51969 or at 9212357123. Format for this is "KISAAN GOV STOP" A confirmation message to the effect would be sent to the farmer.

Sending SMS using the Portal: On selecting the menu 'Send SMS' the user will be navigated to SMS Creation/ Transmission page. Three options are given for sending the SMS. Enter Mobile Number, Enter Farmers' Database, and Upload Excel Sheet. The template for excel sheet is specified in table format as Name, Mobile, State, District, Block. User can select the file using browse button and the data will be linked to the system using the fetch button.

Nature of SMS: Short messages can be sent to either entire sector or selected entities by choosing one of the options as under:-

i. Entire Sector (s). User can select one or more sectors by clicking on this. Option of selecting one or more sectors (Agriculture, Animal Husbandry, Fisheries and Horticulture) or all four sectors by selecting all radio buttons is also provided. SMS will go to the farmers who have opted to receive messages for these sectors.

ii. Advisory for Selected Sector- On selecting option for Selected Entity (ies), a drop down menu for selecting the Sector category will appear. Five different Crops/Activities in a particular sector can be selected. Farmers who have opted for any of these 5 Crops/ Activities will receive the SMS being sent.

Type of Message & Topics: SMSs to be sent to the farmers can be broadly classified into three categories, viz. Information, Services and Scientific Advisories. First two are basically factual in nature and therefore the messages of these types can be sent by simply checking the official records. However, in case of Scientific Advisories, it has to be ensured that there is no contradiction among the advisories being sent by different people on the same issue in a given area and the advice being given to the farmers is technically correct. Therefore, database of previous advisories issued on various topics through the SMS Gateways is also being populated in the same SMS Portal. This database is searchable by words/ phrases, by topics across various levels and also for a period.

AGMARKNET Portal

Agricultural Marketing Information Network (AGMARKNET) was launched in March 2000 by the Union Ministry of Agriculture. AGMARKNET is a farmer-centric web portal sponsored by the Directorate of Marketing and Inspection (DMI), Ministry of Agriculture and executed by the Agricultural Informatics Division of NIC for linking all important agricultural produce markets in the country, the State Agricultural marketing Boards and directorates and DMI for effective information exchange. A step towards globalization of Indian Agriculture, the NICNET based AGMARKNET facilitates generation and transmission of prices and arrival information from agricultural produce markets and its web-based dissemination for the use of individual producers, farmers organizations, consumers, traders, exporters, NGOs, communication agencies like Radio, Newspapers, TVs, websites, food processing industries, Chambers of Commerce, policy makers, etc. (www.agmarknet.nic.in). IFFCO, a cooperative sector company, has been involved in disseminating the price information to a large network of farmer's cooperatives. Public sector banks, commodity exchanges, news papers, etc. are using the site for publishing the data on prices of agricultural commodities apart from other international organizations.

The AGMARKNET web portal (www.agmarknet.nic.in) is a Government to Citizen, Government to Business and Business to Business (G2C, G2B and B2B) e-governance portal that caters to the needs of various stakeholders such as farmers, industry, policy makers and academic institutions by providing agricultural marketing related information from a single window. The main users of AGMARKNET web portal are producers, traders, exporters, Commodity Boards / Departments etc. The linkages are also provided to Kisan, Krishi World, Indian Agro-net, Agri-watch, Indian Dairy Industry, Agro-India, Forwards Markets Commission, National Multi Exchange of India Limited and NCDEX through the portal to form a close user-group. The portal has helped to reach farmers who do not have sufficient resources to get adequate market information. It facilitates web-based information flow, of the daily arrivals and prices of commodities in the agricultural produce markets spread across the country. The data transmitted from all the markets is available on the AGMARKNET portal in 8 regional languages and English. It displays commodity-wise, variety-wise daily prices and arrivals information from all wholesale markets. Various types of reports can be viewed including trend reports for prices and arrivals for important commodities. Currently in April, 2014 about 3200 markets are connected and work is in progress for another markets. The AGMARKNET portal now has a database of about 300 commodities and 2000 varieties.

How Does Portal Work?

The portal is run by the Directorate of Marketing & Inspection with the technical assistance of National Informatics Centre (NIC) as consultant on turn-key basis. The computer facility has been provided at different markets (called Mandi) at grass root level and connected to internet with DMI Hqrs. An individual market centre enters the data in the prescribed format, using the customized application software "Agmark". The information is either sent through State Government or directly to DMI Hqrs. Faridabad.

From Faridabad it is uploaded to the portal. The data from DMI Hqrs. is transferred to portal twice in a day, first at 1100 hrs. and then at 1600 hrs. except on market/public holidays. Thus the information becomes accessible to the farmers and other end users.

Future development involves linking all the agricultural wholesale markets in the country and establishing strategic alliances with government and non-government organisations to disseminate information to the farmers who operate in these markets. The database developed under AGMARKNET would also be linked to other agricultural databases, for instance, on area, production, yield of crops, land use, cost of cultivation, agriculture exports and imports, and so on, to evolve a data warehouse. This would provide a sound base for planning demand-driven agriculture production. AGMARKNET is also expected to play a crucial role in enabling e-commerce in agricultural marketing. This portal helps in reducing the information asymmetry in agricultural prices and thus is of immense use to stakeholders.

Services offered by AGMARKNET

1. Prices and arrivals daily max, min, modal prices, MSP; weekly/ monthly prices/arrivals trends; future prices from 3 national commodity exchanges.
2. Information about grades and standards for agriculture produces; CODEX standards of agricultural commodities, markets profile, etc.
3. Production and marketing technology (Commodity Profile) of various crops like Paddy/ Rice, Bengal Gram, Rapeseed- Mustard, Red Gram, Soybean, Wheat, Groundnut, Sunflower, Black Gram, Sesame, Green Gram, Potato, Maize, Jowar, Cotton, Grapes, Chilies, Mandarin Orange etc.)
4. Market Profiles (Contact details, rail/ road connectivity, market charges, infrastructure facilities, revenue etc.)
5. Other Reports (Best Marketing Practices, Market Directory, Scheme Guidelines, DPRs of Terminal Markets etc.)
6. Research Studies on agriculture marketing issues.
7. Information about companies involved in contract farming.
8. Schemes of DMI for strengthening agricultural marketing infrastructure.

aAQUA (almost All Questions Answered)

aAQUA which stands for almost All Questions Answered is a farmer-expert question & answer database supporting Indian languages. It is an online multilingual, multimedia agricultural portal for disseminating information from and to the grassroots of the Indian agricultural community. The technology for almost All Questions Answered (aAQUA) was developed by Developmental Informatics Lab, KReSIT, IIT Bombay and was sponsored by Media Lab Asia and India Development Gateway (InDG) Foundation's research & development center. aAQUA simultaneously addresses two major challenges in farmer outreach programs viz geographic reach and customized delivery. It answers farmers queries based on the location, season, crop and other information provided by farmers. An aAQUA question is posted either by a registered user directly or through a telecenter / kiosk operator who has an account in aAQUA. Usually the question is from a farmer whose profile information provides details such as crop, farm size, pesticides and fertilizers used dosage etc. The prices of various commodities along with their varieties are displayed spatially over a map. The user can decide where to sell his produce to get the maximum profit, depending on the prices and the distance of the markets.

aAQUA makes use of novel database systems and information retrieval techniques like intelligent caching, offline access with intermittent synchronization, semantic-based search, etc. Agricultural content repositories (Digital Library), Agri-price information (Bhav Puchiye), farmer schemes and various operations support databases (aAQUA: Q&A) have also emerged from the experience of aAQUA deployments. aAQUA's large scale deployment provides avenues for researchers to contribute in the areas of knowledge management, cross-lingual information retrieval, and providing accessible content for rural populations. Apart from agriculture, aAQUA can be configured and customized for expert advice over mobile networks and the internet in education, healthcare and other domains of interest to a developing population. aAQUA is being spread geographically by building strategic partnerships with the state governments, kiosk network providers and agricultural expert organizations. Additional services being considered for aAQUA as part of the scaling up effort include (i) weather reports (ii) database of populous villages and their location information, (iii) quality standards for exports. Apart from agriculture, aAQUA, Bhav Puchiye and the Digital Library can be configured and customized for expert advice and content in education, healthcare, e-Governance esp. by organizations who are working in connecting laymen with experts.

Agropedia

Agropedia is a vision which seeks to address lacunae in Indian agriculture knowledge and application, specifically the lack of content, organized information, and extension services which are serious challenges. It has developed delivery mechanisms such as vKVK and KVK-Net, which are a means to connect extension scientists and farmers, based on its web platform which hosts information on agriculture and rural livelihood. The web platform, agropedia, is unique in the agricultural domain as it is semantically organized and enabled so as to assist one in getting exactly what information or service she or he is looking for. Along with content in the library section, certified by the Indian Council of Agricultural Research (gyan dhara), it also allows space for interaction, knowledge co-creation (jana gyan) and social networking (chat, blog, forum) making the portal dynamic and learning participative. The objective of agropedia is to build up an agricultural knowledge repository for India in the form of textual, audio and video clippings, which is delivered through the web. It wants to make a proper knowledge network among the different actors of the agriculture community by developing an open platform of knowledge sharing in the agricultural domain. The combination of getting authentic agricultural knowledge and information with the option of sharing own views can hardly be found in any Indian agricultural website. In this sense the process of gathering the wide spread agricultural knowledge and dissemination of this knowledge through e-network is unique. Agropedia provides the option for more specific search on a specific agricultural topic. In the agropedia one can increase his knowledge as well as can share his agricultural information with others. 'Extension Material' includes different types of content, provided or thoroughly checked by the agricultural scientists. Beside this, anyone can create content and share his opinion and experiences through wiki, blogs, forum and online chat. Facilities provided in this website are incorporated by taking into account the requirement of different types of users. Everyone can find and use his favourable way for creation and sharing of agricultural knowledge as the contents are provided through different form.

Agropedia basically has two types of search: substring (basic search just like google) and the other one is semantic (tag specific-based on taxonomy). Now the search we see on the right top page of agropedia is a substring search: If you type rice fertilizers there, it will give some 4 pages (approx 40) results. But if you search it inside the section Library of agropedia which has fully semantic based searching, there we need to type the keyword which matches completely with tag set which are present in taxonomy. To get the exact keyword first we need to click on either browse search interface of Library page or Content Index search interface using these interfaces we get the exact "word" to be searched. Example: After selecting Rice in the first box we should write "basmati rice" which is a tag in tag set and not just "basmati".

In agropedia only the consortium partners have the authority to upload content in the Extension Material. The University of Agricultural Sciences- Dharwad and Zonal Project Directorate-IV, Kanpur, are the main content providers. IIM Calcutta, IIT-Kanpur and ICRISAT may also provide some content. However, in the Interaction section which includes Agrowiki, Agroblog and Agroforum each and every user can provide content on agricultural related issues. Contributions to the Interaction section are highly encouraged. All the registered users can add content under all the categories. Registration is free and any one can register just by creating an account in agropedia portal. Other supporting partners include Food and Agricultural Organization (FAO), Consultative Group on International Agricultural Research (CGIAR), Indian Council of Agricultural Research (ICAR), many ICAR organizations, National Agricultural Research System (NARS), IIT- Madras, ICRISAT, Hyderabad, IIM Calcutta, UAS-Raichur, ZPD-IV, Kanpur. Some NGOs like WIZMIN, Shramik Bharti, Central Research Institutes, State Agricultural Universities like Acharya N G Ranga Agricultural university (ANGRAU) Andhra Pradesh, and other agricultural related institutes as National Remote Sensing Agency (NRSA), Sri Aurobindo Institute of Rural Development (SAIRD)- Andhra Pradesh (Gaddipalli), TATA-Chemicals. Moreover, some consortium partners are linked with two Krishi Vigyan Kendras (KVKs)/ Agricultural Research Stations (ARSs) each.

The portal is very useful to the extension workers from KVKs and NGOs, agricultural researchers, scientists, experts, academic institutions, organizations, Agricultural Research Stations (ARSs), advanced farmers, traders, retailers, and self-help-group. Anyone with a keen interest in agriculture can access this site, contribute towards it and engage in healthy discussion.

vKVK: vKVK is a platform that connects KVKs with farmers through internet and mobile technology. The objective here is to bridge the gap between the farmers and the KVK expert. vKVK is developed by IIT-Kanpur in a consortium mode as part of ICAR-NAIP supported project entitled "Engaging Farmers, Enriching Knowledge - Agropedia Phase II" vKVK service started in August, 2011 as a pilot project in four states with the official partners viz International Crops Research Institute for the Semi-Arid Tropics (ICRISAT)-Hyderabad, University of Agricultural Sciences-Dharwad and ZPD-IV. Initially, the service was started in 24 KVKs and is now being up-scaled in 190 KVKs covering more than 30,000 farmers. The vKVK service is cutting across all the service providers and can be accessed flawlessly on even low end mobile handsets. All vKVK services are free of cost to the farmers.

KVK-Net: Krishi Vigyan Knowledge Network is a service to integrate all the KVK scientists and enhance interaction, coordination and cooperation. It is a platform for them to share their experiences, discuss issues and enrich each other in the process.

AgRIS

Agricultural Resources Information System (http://agris.nic.in) is a step towards establishing a location specific e-Government Model for the Resource-Poor-Farmers. The Agricultural Resources Information System (AgRIS) is the Central Sector Scheme for Strengthening/ Promoting Agricultural Information System in the Department of Agriculture & Cooperation (DAC), Ministry of Agriculture, and Government of India. AgRIS is "A Needed Domestic Strategy for Sustainable Agricultural Production and Sustainable Rural Livelihoods in India". It is a e-Government Programme for fostering agricultural growth, poverty reduction and sustainable resource use in India at grassroots level and also "A Step towards establishing a location-specific e-Government model for the Poor". The implementation of component AgRIS will facilitate development of typology specific agriculture development plan in the country. This Project is being executed by Agricultural Informatics Division of National Informatics Centre.

Objectives of AgRIS

- "Development of Decision Support Systems (DSS) on Production Practices and Systems". The DSS proposed under AgRIS will facilitate farmers in adopting appropriate agricultural production practices.
- "Creation of Metadata" to become the Country's initiative of "National Spatial Data Infrastructure (NSDI) on Agriculture".
- Preparation of Guidelines on standardized methodology/ best practices to be used for building Agricultural Resource Information System in similar districts of the Country.
- Develop District specific Database on Agricultural Resources and system to manage the information.
- Capacity Building through Human Resource Development for associated officials in the districts.

AgRISNet

Agriculture Resources Information System Network (AGRISNET) is a Mission Mode Project under National e-Governance Plan of Government of India, Department of Agriculture & Co-operation. Ministry of Agriculture has decided to launch a Central Sector Scheme titled, "Strengthening / Promoting Agricultural Informatics & Communications" of which one component is AGRISNET. The portal provides the end users with the ability to access services through an online platform at their door-step. It is an Infrastructure network up to block level agricultural offices facilitating agricultural extension services and agribusiness activities to usher in rural prosperity. Under this scheme most of the State Governments have established information rich agricultural websites. For example Sikkim AGRISNET (http://www.sikkimagrisnet.org), Andhra Pradesh Agri-portal, (http://www.apagrisnet.gov.in), Uttar Pradesh AGRISNET Knowledge Portal (http://agriculture.up.nic.in), Tamil Nadu AGRISNET portal (www.tnagrisnet.tn.gov.in) and so on.

ARISNET (Agricultural Research Information System Network)

Indian Council of Agricultural Research is an apex body under Indian National Agricultural Research System. Therefore, all organisations under NARS were networked by ARISNET. Agricultural Research Information System Network envisages strengthening agricultural research, education and extension services through networking and improved management of information. ARISNET connects 4 National Research Bureaus, 48 Central Research Institutes, 11 Project Directorates, 30 National Research Centres, 160 Regional Research Stations, 90 All India Coordinated Research Projects, 01 Central Agricultural University, 60 State Agricultural Universities & their Colleges, 120 Zonal Research Stations, and 630 Krishi Vigyan Kendras. Over the years, 237 ARIS Cells have been established. ARISNET use VSATs, leased line, RF for WAN and LAN at the institute campus. ARISNET envisaged bringing Agricultural Research, Education & Extension Information and Resources together for improving quality of technology & services to farming communities & other stakeholders. The main objectives are: Creation of central repository of scientists and projects undertaken by them at micro level; Optimization of resources through better planning, prioritization, coordination, reducing duplication of research, greater use of information for manpower planning.

DACNET

Department of Agriculture and Cooperation Network is NICNET based e-governance project executed by NIC to facilitate dissemination and exchange of faster and reliable agricultural information across the Ministry of Agriculture and Directorates and Field Units of Department of Agriculture and Cooperation (DAC). It is aimed at qualitative enhancements of work culture through better transparency, streamlining of existing administrative and technical methods and practices, greater integration and use of the data collected from variety of data sources, moving towards a paperless office environment, providing telecommuting features and better communication through email for better coordination, knowledge sharing and research. DACNET supports decision making for planners, integrates Government-to-Government (G2G) functions, connects Agri-Business Partners (B2B) and Farmers (C2C) and ensures IT empowerment of the officials of the directorate. (Source: http://dacnet.nic.in)

HORTNET (Horticultural Informatics Network)

The Government is releasing thousands of crores of rupees under National Horticulture Mission (NHM) since 2005 to the States and Union Territories of India, to promote holistic growth of the horticulture sector covering fruits, spices, flowers, aromatic plants, cashew and cocoa through a variety of interventions relating to production and productivity improvement, HRD, post-harvest management and marketing covering more than 60 components. 372 Districts are covered under NHM. Besides, 13 National Level Agencies are also involved. The real beneficiaries' data is not easily available at all the required levels of decision making and monitoring. The transparency, authentication, verification, validation, timely reporting, coordination, communication, performance management, availability of required information in time, rational and budgeting, planning, avoiding duplication efforts are the several challenges. To address these challenges, and provide all the required services to the farmers' online HortNet project was designed and developed in September 2009.

- **Uniqueness:** HORTNET project is a unique intervention to accomplish e-Governance in NHM wherein total transparency has been envisaged in all the processes of workflow i.e. online application filing, authentication, processing and online payment to the beneficiary's bank account. SMS based e-alerting is an added feature of the system.
- **Functionality:** For the first time in the country, a single registration facility for all the State and Central Govt schemes by the farmers has been created besides all the required services, such as information about the welfare schemes, acceptance of applications online, acknowledgement, counselling, selection, releasing for funds electronically, alerts about the pests, diseases management besides best practices through the mobiles so as to reach the unreached.
- **Stakeholders:** Farmers, officials of the State Horticulture Mission and its allied sectors, other government officials, auditors, consultants, inputs/ fertiliser supplying agencies and its distributors, knowledge workers and researchers. HORTNET enables all stakeholders to gain access to information and data on real time basis for ensuring better delivery mechanism and efficient administration of the scheme.
- **Objectives**: There is no comprehensive integrated single and simple transparent, efficient and effective system to manage farmer welfare

horticulture schemes in the country. Only statistics & no individual beneficiary details are available at the State and National level. Farmers are not getting timely advise as a result they incurring losses. Fund flow management, budgeting, planning used to be laborious and time consuming and error prone activities.

a) To simplify and seamlessly integrate the workflow among Ministry, Govt of India, State Govt. functionaries located at State, District, Sub-division and village level.

b) To bring in efficiency and transparency in managing the farmer beneficiary welfare schemes of the government.

c) To provide one-stop, non-stop solution to benefit government officials, banks, farmers, extension workers, researchers, belonging to various departments.

d) To introduce state of the art least cost ICT technology based scientific workflow and process management system.

e) To verify online the land details provided by the farmers.

f) To ensure proper utilization of all plan funds through a transparent workflow system using HortNet.

Innovativeness of the Project

a) For the first time in the country at least cost, Internet, Mobile, GIS, GPRS and GPS technologies have been harnessed to reach the un-reached and ensure transparency.

b) Integration with Google maps for providing spatial view.

c) Very user friendly and interactive.

d) Farmers or their well wishers or the Agriculture/ Horticulture extension worker can apply online.

e) Application gets processed transparently. Benefits get delivered to the farmers.

f) This is the first cost effective solution to the entire workflow automation to render services Government to farmers. No other farmer centric Govt portal has this many features and services profile.

g) HortNet is unique Farmers centric application and a SAAS model.

h) Online Scrutiny and approvals. Role and location based access control for concerned officials. As a result, the work which use to take several man months, is now hardly taking few minutes.

i) Third Party field verification and auditing and reporting using mobile, GIS, GPRS and multi-media technology to check frauds very easily which hitherto impossible.

j) User friendly, local language complaint, so existing government got training over Video Conferencing from respective district head quarters and started implementing the project themselves. For the first time in the country such a low-cost roll-out method adopted for sustainable and successful e-Governance. (*Source*: http://hortnet.gov.in)

SeedNet India Portal

Seed is the critical and basic input on which performance and efficacy of other inputs depends for attaining sustainable growth in agricultural production. Ninth Five Year Plan has recommended establishment of "National Seed Grid" (NSG) as on-line computer based network linking the National and State level Seed corporations, to serve as database on availability, supply and demand for different varieties of seeds in different areas, facilitating the optimum utilization of available seeds in the country. Development of Seed Informatics is one of the ICT initiative and a part of e-Governance implementation in the Department of Agriculture and Cooperation.

SeedNet: Portal is a "One-Stop Service" for all information related to Seeds. It integrates all seed players in the country. All the schemes launched by the Government of India are available online. SeedNet is Seed Informatics Network under ministry of Agriculture, Government of India initiated in 2008. SeedNet India Portal Striving for prosperity of farmers is a national Initiative for information on quality seeds. It provides information about the requirement, production and availability of Quality Seeds at different State Agricultural Universities and Breeder seed producers also the availability of Foundation and Certified Seeds at different locations. All relevant government orders, Acts/ Control Orders, Rules, Quality Control measures, details about extension services, data on export and import of seeds and programmes and schemes of the DAC have been provided on the Portal. The salient features of over 3500 varieties, notified/released are available online from 1966 till 2008. The complete Seed Supply Chain from Indent to Allocation is available online. Different agencies can provide and update requisite information. Various users, particularly farmers will be able to access the information from a single source. Its outreach is all over India and the stake holders are Department of agriculture and Cooperation, Government of India, ICAR, State Agriculture Departments, SAUs, NSC, SFCI, SSCs, SCAs, STLs, seed Producers, seed Processors, Dealers, Private seed organisations, Researchers and Farming Community.

Services offered by Portal: SeedNet India Portal provides G2G, G2C, G2B, G2F services with scope for B2B service also. The following online web enabled applications are available on the portal.

- Seed Varietals Information System.
- Seed Management Information System.
- Policy decision on Export and Import of Seeds.
- Seed Certification Management System.
- Management Information System for Seed Testing.
- Registration of Seed Dealers.

All relevant processes and procedures of seed certification, seed testing, seed processing etc. along with manuals have been made available online. Databases have been created to facilitate the above listed online applications. The content available on the portal has been provided by the DAC. It is a one stop single window service for all matters related to seed sector. The applications are easy to use and are user friendly with proper documentation. The SeedNet India portal and applications are hosted on the central server at Internet Data Centre, NIC, New Delhi with full back up support, disaster recovery, high bandwidth, firewall and 24x7 availability. (*Source*: http://www.seednet.gov.in)

E-KRISHI KIRAN

The project was established and implemented in the Gujarat state by Anand Agricultural University (AAU). The program e-krishi kiran or e-Soil Health Card Programme can be accessed by http://shc.gujarat.gov.in . e-krishi kiran program, is a attempt to promote scientific agricultural technologies by enabling the farmers to get required specific information through use of Information Technology. Government of Gujarat has thought out a new way by launching e-krishi kiran program which is an online program of technology transfer with an individual farm condition in focus. This program is expected to bridge the distance between Scientist – Extension workers – Farmers and Input Dealers effectively. It helps making transfer of technology more scientific, precise, easy, and need based. The e-krishi kiran program is a web based information system designed to run on a networked environment including intranet, internet and on GSWAN (Gujarat State Wide Area Network). This is a repository of agricultural information for the benefit of farmers, agricultural scientists and decision makers. This e-Extension program is a unique information initiative of its kind in India pioneered and initiated by the Government of Gujarat for the benefit of farmers at the grass-root level. In first phase it was decided to provide 5 lakh soil health cards to farmers every year, thus covering all farmers within five years.

Objectives of e-krishi kiran

- Direct advice to individual farmer; a new approach in extension.
- Direct access of agricultural scientist at village level to bridge gap between actual and potential production.
- Increasing the income of farmers.
- Analysis of soil of all the villages of the state.
- Recommendation for fertilizer usages for each farm, based on soil analysis.
- Suggest alternate cropping pattern based on moisture availability index.
- Provide Soil Health Card to each farmer to guide him about his soil conditions and cropping practices to be followed.
- Promoting optimum fertilizer use and fertilizer-use-efficiency through soil test method by analyzing soil of every farmer's field.
- Developing a crop planning for the rainfed areas, which register low and uncertain productivity and remain major reason for poverty among the farmers.
- More elaborate data base, their quick utilization and networking to enhance efficient extension services at various levels of Department of Agriculture and Cooperation (DAC) viz. village, takula, district and state level offices.
- Better crop production.
- Better utilization of knowledge and research through the dissemination of information using various avenues of Information Technology.

How Does it Work?

The main objective of e-krishi kiran program is to promote scientifically sound agriculture. For example, for achieving scientific fertilizer use, soil samples of individual farmers were collected, analyzed in the laboratories and on the basis of available nutrients present in the soil, fertilizer rate for the crop that the farmer grows was recommended. Similarly based on soil moisture index based approach, suggestions for more profitable rain fed crops were made.

Under this program, AAU has developed the web-base application software, which generates and provides the fertilizer recommendations on the basis of soil analysis and the nutrient requirements of the crop for each field. This will increase the efficiency of the fertilizer and saves consumption of the fertilizer. Inputs given to the system includes: Farmers record with their Name, District, Village, Survey number, Soil Health Card (SHC) number, etc.; Soil fertility data based on soil test report regarding organic carbon, available phosphorus and potash, EC and pH; Soil type and depth; Taluka/ Block wise rainfall data; Agro climatic zone wise crop cultivation practices; Recommended NPK doses; Alternative crop planning; Crop economics data; Action plan information (information of Taluka / Village action plan).

On Basis of Above Input Data Following Output can be Generate by the system

- Recommendation of the fertilizers needed for a particular crop based on a nutrition status of farmer's soil on the basis of low, medium, high soil fertility rating.
- The soil fertility index of each village on the basis of available soil analysis reports of the insisting farmers, to generate village soil health card for the use of other farmers who had not made soil testing of their fields.
- Recommendation of the possible alternative crops to a farmer for better crop production based on his cropping practice and weather condition in his area considering the facts like moisture availability index, available water capacity, length of growing period, surplus water and supplementary irrigation.
- Recommendation of the alternative crops with a generic ranking of crops in terms of profitability, looking into the various costs involved for producing that crop.
- The state wise, district wise, taluka wise and individual farmer wise model action plans for crop production.
- The various FAQ's for farmers about agriculture and animal husbandry in a well classified manner.

The above personalized information and recommendations enable farmers to take right decisions about their farms.

Digital Green: Participatory Videos for Agricultural Extension

Digital Green work with existing people based extension systems, aiming to amplify their effectiveness through ICT enabled approach. Its model combines technology and social organization to maximize the potential of capacity building community members on improved, sustainable agriculture, livelihood and health interventions. It also facilitate knowledge exchange between community engagement for partners looking to learn, contribute and connect on social innovation practices toward improving lives in rural communities. Digital Green works with partners throughout the entire experience to share knowledge and capture feedback with supported technologies that allow partners to locally produce and share videos in villages all around the world. Within four years, digital green have reached 2,000 villages across India, Ethiopia and Ghana and produced more than 2,600 videos in 20 different languages; Improved lives of more than 160,000 community members (70% women) in South Asia and Sub-Saharan Africa; Increased the cost-effectiveness of development efforts by 10 times. Digital Green aims to reach 10,000 villages in India and Sub Saharan Africa, enabling the communities to live with dignity.

How does Digital Green Work?

- Village level mediators, trained by DG and partners, produce and share videos on locally relevant agronomic, health and livelihood practices to motivate and educate community members. The equipment used for video production and dissemination are low cost, durable and easy to use, and adaptive to diverse environments.
- A community video production team of four to six individuals in each district creates videos, averaging 8 to 10 minutes in length, which are screened for small community groups twice a week using battery-operated pico projectors.
- A facilitator from the community mediates a discussion around the video screenings by pausing, rewinding, asking questions, and responding to feedback.
- Regular adoption verification visits are scheduled for gauging the impact of the dissemination on actual practices.

Key Principles of Digital Green

- Partnership: Focus on developing and delivering digital content to improve the cost effectiveness and efficacy of organizations involved in agricultural and livelihood research and extension.

- Community ownership: Adoption of knowledge exchange model by communities for their improved wellbeing by enabling them to be content producers and consumers, and empower communities to have a voice.
- Localization: A human centered and contextual design approach which helps in successfully customizing to suit diverse contexts and stakeholders.
- Cost realism: Operate in environments with limited infrastructure and financial resources, which entails paying attention to costs at all levels, and limiting the investment in expensive equipment.
- Mediation: Videotaped demonstrations are not a complete extension solution since they lack the interactivity expected of good extension. Therefore, a local facilitator, who occasionally pauses or repeats video to engage the audience with discussion and capture feedback.
- Training: Capacity building of partners, extension staff, and village facilitators in the social aspects of the approach for effective delivery of key messages.
- Gender: Targets for participation of women in the intervention, for instance as local facilitators, are determined and tracked. Particular attention is paid to the gender balance in recruiting women for video production. Topics of relevance to women are included in the content, especially topics related to crops grown by women.

e-Krishi Vipanan

Regulated agricultural markets or 'mandis' have revolutionized the system of agricultural marketing in Madhya Pradesh. The days have gone when a farmer used to come to the mandi and was forced to sell his crop even when he knew that the crop was not being valued correctly. Now a farmer is empowered with the information that enables him to make informed decisions as to when and where to sell. In other words, now he can decide on the mandi where he wants to sell his crop, even demand more prices in the same mandi, or refuse to sell at all if not given the right price. All this has been made possible by the unique ICT initiative called e-Agricultural Marketing or e-Krishi Vipanan (EKVI). e-Agriculture Marketing project of Madhya Pradesh Government was conceived and executed by Madhya Pradesh Agricultural Marketing Board (Mandi Board) and Madhya Pradesh Agency for Promotion of Information Technology (MAP_IT), on Build, Own, and Operate basis. The project was implemented on PPP (Public Private Participation) mode by a consortium of vendors wherein the vendors made operations effective and transparent by collecting and disseminating real time online information and help the concerned stakeholders in effective decision making which will eventually lead to grain less mandis. EKVI Project involves use of ICT for automation of mandi board Head Office, seven regional offices, and 231 Mandis and their associated Sub-market yards and Nakas (Inter-state border check posts) across the state of Madhya Pradesh. Most of the Mandis and Sub-mandis are located in villages having 6 million Farmers with 70,000 licensed traders. The data generated at mandis with regard to agricultural produce, sale, etc. is captured 'online' through smart card terminals, transferred to computers in mandis, and transmitted on a communication network to the associated regional office and Head Office via V-SAT. This information is then accessible at specified nakas and a few other points for verification of documents on a 24 x 7 basis. To augment agricultural marketing in the state and ensure that transactions are transparent, prompt, farmer and trader friendly, the Mandi Board desired to computerise mandi activities through the networking of mandis and the Mandi Board to eventually achieve true e-Agricultural Marketing in Madhya Pradesh. Due to complex activities and huge volume of commodities it was not possible to have total transparency and effectiveness without automation of the activities. Therefore, it was felt that automation and networking of mandis would not only improve the effectiveness of trading but also benefit the farmers, traders and the government in collection, processing, analysis and dissemination of information, policy making and revenue collections.

During the pilot project and in the initial roll out of the project had been substantial increase in the collections of mandi fees going to the government. The rates of various produces in the neighboring mandis and the National mandis are displayed in the mandi display system through the information collected (again from different mandis) by the project itself and through scanning of available information with sources such as the National Commodity Exchange (NCDEX), among others. All these are helping the farmers to reach at conscious informed decisions to sell their produce. The system is also helping in quick issue of transit permits (for transportation of agricultural produce) and its reconciliation at various Interstate check-posts (Nakas), thus ensuring that the agricultural produce is not transported without a valid permit. All financial transactions are reconciled with banks on a day-to-day basis. The entire essential 'management information' is made available to the management of the mandis, regional offices and HO of the Mandi Board periodically as specified. To benefit the Indian farmers the ICT initiatives like e-Krishi Vipnan (grain less mandies) can be replicated in other states also.

e-Sagu (e-Cultivation)

Word 'e' stands for electronic and 'Sagu' is a Telegu language word which means cultivation. Thus e-sagu means cultivation with the help of electronic and digital means of communication. International Institute of Information Technology (IIIT), Hyderabad developed e-Sagu system for bridging up the information gap existed between agricultural experts and the farmer. It is an IT based personalized agricultural extension system to improve agricultural productivity by disseminating improved agricultural advice to the farmers, both in a timely and personalized manner. e-Sagu is a ICT tool for delivering personalized Agro-advisory services at the door-step of the farmer. The advice is provided on a regular basis (weekly) from sowing to harvesting which reduces the cost of cultivation and increases the farm productivity as well as quality of agriculture commodities. In e-Sagu, the agricultural experts generate the expert advice based on the information about the crop situation received in the form of both text and digital photographs. These photographs are sent by some educated and experienced farmers (called coordinator) in the village. e-Sagu is using the power of IT to improve the performance of agricultural extension services. e-Sagu was started during Kharif season of 2004. The e-Sagu system was implemented by delivering advisory to 1051 cotton farms in three villages of Warangal district in Andhra Pradesh. A team of agricultural experts has stayed at IIIT, Hyderabad (India) and delivered 20,000 pieces of agricultural expert advice to 1051 cotton farms of three villages of Warangal (AP) by looking at digital photographs and other farm information supplied by coordinators in these villages. The project was successfully implemented (Reddy et al. October 2005, Reddy et al, December 2005). During 2005-06, as a scaled-up experiment, e-Sagu system was implemented for cotton, chili, rice, groundnut, castor, red-gram and other crops for 4894 farms in 35 villages spread over six districts in Andhra Pradesh. In addition, the e-Sagu system has evolved to deliver advisory service to fish-farmers. About 160 fish ponds were covered during the previous year. By the end of 2005-06, there were 8 e-Sagu local centers which had delivered 35,925 advices covering 30 crops and fishing rearing. About 2.6 lakh crop photographs and 8918 fish photographs were taken in the process. As of now, the turnaround time for advice delivery is between 24 to 36 hours.

The farmers are happy with the expert advicc as it is helping them to improve input efficiency through the use of integrated pest management (IPM), judicious use of pesticides and fertilizers etc. At the same time benefits accrue through higher yield.

How does e-Sagu work?

In e-Sagu, instead of agricultural expert visiting the crop, the crop situation is brought to the agricultural expert using both text and digital photographs. A team of agricultural experts work at the e-Sagu main lab, supported by an agricultural information system. One e-Sagu local center consists of a few computers and a computer operator and covers a group of about ten villages. Educated and experienced farmers work as coordinators. Depending on the crop, each coordinator is assigned a fixed number of farmers. He collects the registration details of the farms including soil data, water resources and capital availability and sends the information to the main e-Sagu system. Every day the coordinator visits a fixed number of farms and takes 4-5 photographs for each farm. A CD is prepared with the photographs and other information and transported to the main system by a regular postal service. The agricultural experts with diverse backgrounds (Entomology, Plant Pathology, Agronomy etc.) at the e-Sagu main lab analyze the crop situation with respect to soil, weather and other agronomic practices and prepare a farm specific advice. At the local e-Sagu center, the advice is downloaded electronically through Internet. The coordinator collects the advice prints out and delivers it to the concerned farmer. The coordinator then explains the advice to the farmer gets the feedback and sends it to agricultural expert. In this way each farm gets the proactive advice at regular intervals starting from pre sowing operations to post harvest precautions. Some of the advantages of the system over existing traditional system of extension are availability of a team of diversified experts at a single place, saving of time, money and energy, enabling correct diagnosis of the problem, strong database to support decision-making, zooming facility adds an extra dimension, accountability to the farmer, capacitating of rural livelihoods and employment generation, documentation of success stories and content development, feedback helps to evaluate and improve the performance etc.

IFFCO Kisan Sanchar Limited (IKSL)

Telecommunications is a fast growing arena to transform Indian rural landscape. In the recent years, it has demonstrated its potential to play a vital role in contributing to the empowerment of people living in Indian villages. The Indian Farmers Fertilizer Cooperative Limited (IFFCO) launched a project to leverage information and communication technology, in particular mobile telephony to disseminate expert agricultural knowledge to small-scale farmers in rural India. IKSL is a joint venture between IFFCO and Airtel. In India, farmers in rural areas have little access of knowledge about improved farming practices and rural development services. IKSL manages the distribution of mobile phone's SIM cards, which is used for communication and the same is turned into a powerhouse of knowledge for empowering people living in villages through relevant & pertinent information which is being provided by IKSL through Value Added Service (VAS). IKSL services are available in 106 zones of 18 states. To provide efficient and customized services 57 content managers and 45 expert panels are working as team.

Value Added Service of IKSL

1. **Free Voice Messages:** Through the Green SIM Card, every day, five free voice messages are delivered to the subscribers. Each such voice message is of one minute duration and cover diverse areas like soil management, crop management, dairy and animal husbandry management, horticulture and vegetable management, plant protection, market rates, weather forecasts information, human & cattle health, employment opportunities, government schemes, etc. These voice messages are prepared by experts on subjects of immediate interest to the rural subscribers. The endeavour is to empower rural subscribers, particularly farmers, with latest information inputs for informed decision making. The country is divided into 60 zones to provide relevant content.
2. **IKSL Helpline:** The subscribers of Green SIM Card of IKSL have the privilege of access to a dedicated helpline service, which has the potential to become a rural lifeline through a 6 digit number 534351. Farmers can get a solution to their problems, queries by using this short code. Experienced professionals are accessible on this helpline, who provide solutions to farmers. Where required, subject matter specialists are taken on conference call to provide a satisfactory resolution to queries.
3. **Call Back Facility:** Every day 5 Free Voice Messages are delivered to subscribers in local language on areas of interest to rural India. These voice messages are delivered through the mobile phone to subscribers. However, for those subscribers, who could not pick up the voice message call or have missed a voice message, facility is available to call a short code which provides for listening to the messages again. This facility can be used during any time of a day.
4. **Other Programmes through Mobile Phone:** Apart from the free voice messages and helpline, IKSL organizes live phone-in-program on specific subjects of interest to farmers. In these 'Phone In' programmes, subject matter specialists join IKSL's officials at one place once in a fortnight for redressal of issues, clarifications, etc. The Helpline number used by farmers is also used for these programmes. Mobile based quizzes are organized to sharpen the knowledge levels of subscribers and attractive

gifts are provided to winners. Mobile based quizzes are also conducted to encourage subscriber participation. (Source: http://www.iksl.in)

KISSAN - Karshaka Information Systems Services and Networking

Karshaka Information Systems Services and Networking (KISSAN) is an integrated, multi-modal delivery of agricultural information system, which provides several dynamic and useful information and advisory services for the farming community across Kerala. It is one of the leading citizen centric e-governance projects of the Department of Agriculture, Govt. of Kerala. The project was conceived, developed and managed by the Indian Institute of Information Technology and Management- Kerala for the Department of Agriculture, Govt. of Kerala.

The basic objective of this project is to provide "Right Information to the Right Person(s) at the Right Time in the Right Place (s) and in the Right Context" dynamically using a combination of advanced technology like Web Technology, Television based mass media programs, Telephone based advisory, Mobile SMS based advisory and broadcast service, dedicated online Agri video channel provides video on demand service etc which, involves effective collaboration of experts from key organizations for effective information delivery and knowledge empowerment on demand seamlessly to all farmers across Kerala.

The key feature of KISSAN is the integrated service delivery model that makes available to the experts from any agriculture related organization any mix or all of the above modes of communications to reach timely and effective assistance to farmers anywhere in the state.

The project solves the problem of content gaps by providing the authentic agricultural information though various delivery methods like Television, Internet, Telephone, and Mobile etc. The farmers can choose any medium to seek the relevant information.

Origin of the Project

The Government of Kerala had appointed a "High Level Committee on IT and BT in Agriculture" under the chairmanship of Prof. V.L Chopra (President National Academy of Agricultural Sciences, and Former Director General, ICAR) to explore the possibilities of Information Technology and Biotechnology for the enhancement of Agriculture in Kerala. Based on the detailed study and importance the committee recommended two major projects each in IT and BT. The project was conceived by Indian Institute of Information Technology and Management- Kerala for the Department of Agriculture. The Project was officially launched on 1st November 2003 (Kerala day).

Major Services of KISSAN

The core deliverables and achievements of the project is an integrated multi-component, multi-modal delivery of Agriculture Information Services system that is accessible anywhere, anytime by all concerned. The major services of the project are

1. **Online dynamic Agri Advisory Service:** The dynamic portal based online Advisory service (www.kissankerala.net) is a major hit of the project. The portal provides an online platform for the farmers to interact with the expert scientists and agricultural extension officers in an interactive way. The portal also provides several dynamic advisory services like market information, weather and crop advisory, expert system on fertilizer recommendation and so on. Through the online query management system, the experts in the project have answered more than 15000 queries of the farmers. Through the online fertilizer recommendation system, the farmers have generated more than 12000 fertilizer recommendations (in local language) for their preferred crops.

2. **Kissan Krishideepam:** A weekly Agriculture Television program in Malayalam, that provides selective information dissemination of best practices, success stories, departmental news, news on various schemes, market analysis, cultivation methods, analysis of current issues, etc. Care has been taken to ensure that KISSAN Krishideepam is authentic and totally produced in-house by agricultural and media experts. The project produces and telecast a weekly television program (30 minutes duration) and telecast three times over the leading satellite channel (Asianet) in Kerala. The program now reaches to more than 25 lakhs regular viewers across the state and beyond. The project has completed the production and telecast of 294 unbroken weekly episodes during the last 5.5 years.

3. **Online Agri video Channel:** The project has launched the country's first online video channel in Agriculture. More than 100 selected videos (telecast quality) were made available through the channel. This becomes a single window access to all the agriculture and allied topics for continuous learning for the farmers at free of cost. (www.youtube.com/kissankerala).

4. **Tele-Advisory Services:** The project also provides telephone based Agri advisory services through a dedicated telephone number (0471-2700965) for the farmers. The farmers can ask any questions to the agricultural scientists and seek expert advice for their crops. As part of this service, the project has developed an extensive crop database across the state to provide location specific advisory services.

5. The SMS based agri advisory services enable the farmers to get the information on very fast and it helps to provide location specific information and alert services

iKisan

iKisan is a comprehensive Agri portal (a one-stop information resource for the Indian farmer) addressing the information, knowledge and business requirements of various players in the Agri arena such as farmers, trade channel partners and Agri input /output companies. The portal was launched by Nagarjuna Fertilisers Company Limited (NFCL). iKisan provides online, detailed content on crops, crop management techniques, fertilizers and pesticides and a host of other agriculture related material. Farmers have a critical need to get timely solutions for protecting and nurturing their crops to get best yields. Addressing this key need, iKisan has developed easy-to-use diagnostic packages for different crops which will be provided on demand. Further, it also provides local agri news, weather and market information to the farmers. It enables farmers to network with other farmers, suppliers and consumers across the world. (Source: www.ikisan.com)

iKisan's Expertise

- ikisan's portal www.ikisan.com is the foremost among the agri portals to address the various needs of the Indian farmer, in the language he understands.
- ikisan is a pioneer in the field of leveraging the internet and IT technologies in the field of agriculture.
- ikisan has an integrated approach that places farmer at its focus and offers total solutions to farmers.
- ikisan has developed both online and offline approaches keeping in view the ground realities of connectivity and developing solutions appropriate to the farmers.
- ikisan has developed innovative interactive models to enable even illiterate farmers to diagnose problem relating to their crops and arrive at suitable solutions.
- ikisan has expertise in conceiving, designing and implementing turn-key projects. A project from Technology Mission for Cotton, Ministry of Textiles, Govt. of India is being implemented along with ECIL. The project envisages developing cotton software for the benefit of farmers, installing and operating the software through computers and touch-screen kiosks

in 250 agricultural market yards spread over 10 cotton growing states in India.

- ikisan has proven experience in developing subscription based farmer service network of kiosks in Andhra Pradesh and Tamil Nadu.
- ikisan, by leveraging on its domain experience in FMS, can take up feasibility study reports for agro products and processing industries / exports.
- ikisan undertakes end to end communication services in the area of farm extension services on a turn-key basis.
- ikisan has entered into a Memorandum of Understanding with the Government of Andhra Pradesh to partner in the Rural E-Seva Project which envisages establishment of around 6,000 information kiosks in rural areas of the state for purposes of e-governance and information dissemination.
- ikisan has been selected by UNCTAD as Tier II Partner for the project "Strategies on Preparedness for Trade and Globalization in India".
- ikisan has the strengths to undertake specialized studies / consultancy assignments in the field of agriculture across the value chain from farms to markets (domestic and export) including GIS utilization.
- ikisan Limited has been invited as a partner in the proposed consortium project on rural livelihoods by Central Research Institute for Dryland Agriculture (CRIDA) under National Agricultural Innovation Project (NAIP).

E-Krishi

E-Krishi, the market driven agricultural initiative through IT enabled agri-business centers in Kerala state addresses the existing gap in agriculture information flow and transaction management. The vision of the project is to establish a well connected farmer's community throughout Kerala who have access to information on market demand, price, good agricultural practices, quality agricultural inputs supported by a technology enabled robust transaction platform that facilitates all their offline activities. The project is piggybacked on the existing resources of Akshaya e-kendras for providing the services. The project was piloted in Malappuram district in 2006 and was funded by UNDP/ NISG. Now it is operational in Kasaragod, Kannur, Kozhikode and Kollam districts with the help of Department of Agriculture and Local Self Government Institutions.

Objectives

1. Preparation of a detailed database of the farmers and businessmen of the state aggregation of responsive farmer community cultivating priority crops as determined by the market demand.
2. Enrolment of buyers in key markets including manufacturers.
3. Enrolment of agricultural input providers (seeds, plantlets, fertilizers, pesticides, technology/methodology providers/consultants, test laboratories and so on) and warehousing facility providers.
4. Enrolment of logistics services support providers
5. Enrolment of banks and insurers
6. Establishment of a robust IT enabled platform where the members can seek information, transact and make or receive electronic payments

How does it work: Sellers (farmers) and buyers (merchants) register through nearest Akshaya Centre or log on to www.e-krishi.org. They post the materials for selling/buying; Post and view advertisements; Virtual meeting of buyer and sellers through the web and effect transaction.

Toll Free Call Centre: Help farmers and buyers in solving trade related doubts. It provides details on good agricultural practices including local weather. The service is managed by qualified agricultural specialists.

Benefits to Member Farmers

1. Access to markets with prevailing price information and ensuring correct price for their products.
2. Access to schemes, subsidies, modern agricultural methods, best practices, soil testing, seeds, plantlets, fertilizers and pest control.
3. Facilities for grading agricultural produces.
4. Logistics support and cost sharing possibilities.
5. Access to micro credits.
6. Agri-insurance support/ faster claim processing.
7. Access to accounting practices and documentation support.

Benefits to Member Agricultural Input Providers

1. Contact farmers producing any specific crop and confirm purchases in advance facilitating better control over supplies.

2. Virtual aggregation of the produces.
3. Buyer/exporters to post their pick-up quantities by date, by market-venue/ warehouse.
4. Plan logistics routes for collection of agri-produce in advance, seeing the offer from farmers in various locations.
5. Select farmers for producing any specific variety of crops on contractual basis integrating supplies of seeds/plantlets, farm maintenance etc.
6. Easy access to customers, facility for systematic campaigning/ demonstrations.
7. Efficient management of schemes/programs e-platform assisting in real time transactions.
8. Searchable directory for all stakeholders with relevant details (contact persons, location, products/services offered, schemes & programs, territorial jurisdiction etc.,) supported by robust platforms for communication, transaction and e-payments.
9. Access to weather information and crop planning. (Source: e-krishi.org)

Agriwatch Portal

Agriwatch portal was started by Indian Agribusiness Systems Pvt. Ltd. (IASL), a company started by a group of young technocrats and trade professionals who have had a long association with the Indian Agribusiness sector. The agribusiness sector is characterized by multi layered channels on the agricultural Input and output side, poor quality of information and analysis about demand, supply, prices, and market trends for various Agri-commodities. The promoters of this organization, coming from agricultural background felt the need to address these anomalies in this sector and therefore have come together to create a professional team and an organization to deliver value to the participants at all the levels in this sector.

Objectives

The Portal primarily aims at filling out the information and communication gap that exists in various sub-sectors of the agricultural economy in general and agricultural commodities trade in particular. For achieving this, the IASL is making use of the latest information and communication technology. The specific objectives are as follows:

- Providing latest and accurate information to various stakeholders in the agribusiness sector such as farmers, Traders, Processors of agricultural outputs, Suppliers of agricultural inputs etc.
- Provide valuable analysis to the trade participants that will enhance their decision taking abilities in trade.
- To enable E-commerce in agricultural products through this vertical portal.

Services Offered by Agriwatch

1. **Customized Agri Commodities Research:** IASL has an information collection network through the country giving us access to over 5 lakh farmers and 25,000 of traders based in all states. Major research services provided include **Value Chain Studies** (Track value addition in a commodity from farm-to-fork), **Sub-Sector Studies** (Assess the structure of a commodity subsector, evaluate the importance of the participants, and study the impact of how the subsector performs in the aggregate), **Consumption Studies** (Study of the downstream industries/ sectors to understand the disappearance of a commodity), **Crop Surveys** (Pre-sowing study of Crop intention, Study of crop progress during the growth phase, Crop harvest estimation based on acreage and yield studies, Production estimates based on surveys, satellite mapping and modeling, Price-forecasts based on time-series modeling).
2. **Advisory and Consulting Services:** Designing and implementing procurement strategies, Timing, locations of procurement, Comparison studies, Supply chain re-engineering, Procurement strategies, Warehousing strategies, Transportation strategies, Risk Mitigation strategies).
3. **Agri Commodities Information Services:** Agriwatch collects data daily in over 250 commodities from over 400 markets in India - One of the largest price-arrival data sets in the country apart from the government sources. The major commodities include:
    - Grains: Rice, Wheat, Maize (corn), Coarse grains.
    - Pulses: Gram, Peas, Moong, Tur, Urad, Lentil.
    - Veg Oils: Soybean, Mustard, Sunflower, Groundnut.
    - Oilseeds: Soybean, Rapeseed/ Mustard, Sunflower, Groundnut.
    - Oilmeals: Soybean, Rapeseed, Groundnut.

- Spices: Red Chilli, Black Pepper, Turmeric, Cumin, Coriander, Cardamom.
- Others: 125 Herbal and Medicinal Plants, Sugar and Jaggery, Cotton, Guar Seed and Guar Gum, Mentha etc.

This market intelligence is available to subscribers on a real-time basis on the web portal in the form of news, market commentaries and spot market price tables and charting. Daily, Weekly and Monthly research reports are also available in the portal to provide more in-depth analysis.

Intelligent Advisory System for Farmers (IASF)

Intelligent Advisory System for Farmers (IASF) is an advisory system for answering queries related to farming activities carried out in North-eastern states of India. The system can be extended with inclusion of any other crops from any State of India. The project covers five major farming activities (Insect Management, Disease Management, Weed Management, Rice Variety Selection and Fertilizer Management) which required expert's advice relating to diagnostic and remedial measures.

IASF aims to improve and strengthen existing agriculture extension services by integrating Information Technology with mobile services. Sharing of knowledge among experts, farmers, students and research scholars are very important to the growth of the agriculture sector. The farmers' queries are stored in a database along with its relevant solution (called CASE) in database. A farmer can ask a question related to the above mentioned farming activities supported by IASF and the system automatically produces a highly probable solution from a large database containing collection of queries and expert opinion given by a team of agriculture experts and subject matter specialist. IASF is also a self learning system that acquires new problems and corresponding solutions. The system can help farmer in critical times where access to an agricultural expert is not forthcoming or due to unavailability of agriculture extension worker in the field. The system is especially useful in States and rural areas where the ratio of extension worker to farmer is very wide.

This system networks registered agricultural experts from various agricultural domain like Plant Protection, Horticulture, Agriculture Extension, Marketing, Agronomy, Soil Science, Seed Technology etc. from the Department of Agriculture, Government of Manipur; Department of Agriculture, Government of Meghalaya; Central Agriculture University (CAU), Imphal; Krishi Vigyan Kendra (KVKs) of Meghalaya in playing an active role as contributor and knowledge engineers for building the knowledge component of the advisory systems.

The main objectives of IASF are

- Provide improved services to the farming community through the use of ICT.
- Advice and help farmers to solve problems related to their farming activities. Otherwise, they need to contact agricultural experts and private extension workers.
- Providing vital and generic information to farmers so that they get periodic alerts on important/useful tips, ideas, knowhow etc.
- Updates farmers on latest technology in Agriculture sector for improved productivity and quality farming.
- Developing an advisory system which can be extended with any other types of crops in any State of India.
- Improving agricultural extension service by using mobile services so that farmers can send queries about their farming problem from their mobile device.
- To develop an educational materials to be used by students for their practical experience with real case scenario.

Beneficiary of the Project

- **Farmer:** Farmers from North East Region of India mainly farmers from Manipur and Meghalaya: Farmers can get direct or indirect advice from agricultural experts for his/her query related to their farming activities covered by IASF. Farmers can also communicate their problems with agricultural experts by sending an SMS text message in a predefined format.
- **Agricultural Expert**: Experts can advise to farmers' query from anywhere using Internet or mobile SMS. Experts can also share knowledge among themselves by suggesting to other expert's advice given to farmers.
- **Department of Agriculture (DOA):** DOA can broadcast periodic alerts and useful information to registered users via mobile SMS and email. Department of Agriculture can use IASF as a tool to train new officers and extension functionaries.
- **Student:** One of major objective of IASF is developing educational materials to be used by students of agriculture for their practical experience with real scenario.

- **Research Scholar:** research scholars will be able to study diversity in farmers' problem across north east region of India and how agricultural experts are suggesting remedial measures on various field problems.

Presently, IASF is benefiting 1886 registered farmers in 6 districts of Meghalaya and 9 districts of Manipur who uses the platform for redressal of their farm problems. 50 Subject Matter Specialist are also registered in the portal to respond to farmer's queries in real time. An agriculture expert can reply to any query from his/her registered mobile phone via SMS by sending his reply to 51969. Farmer/Student can also register themselves to the IASF portal via SMS by sending their name, address in a specified format to 51969. The project can be easily scaled and rolled out with localised content and languages. This is the first of its kind project in North East India, where ICT has come to the rescue of resource poor farmers residing in remote and inaccessible part of the country. The integration of web technology and mobile service delivery gateway has been a boon to this project, where the reach and scale of the project can have a multiplier effect on the agriculture scenario of geographically remote and inaccessible rural areas of India. (Source: http://iasf.cdacmumbai.in/ias/index.jsp)

India Kisan Webportal (www.indiakisan.net)

Food, cloth and shelter are the fundamental needs of the human beings. Agriculture is the one of the important activity for proving food and clothing for the mankind. Due to various revolutions such as Green Revolution, White Revolution, Red Revolution, Yellow Revolution the country is able to produce its food to the present needs as well as immediate future needs. But there are several hunger deaths and farmer suicides along the globe.

Analyzing the problems associated with above in grass root level, an attempt has been made to integrate internet technology with agricultural science to bring the millions of findings of the agriculture scientists developed in the agricultural universities & research institutes within the reach of common man., those who are living in rural areas. A web portal www.indiakisan.net was developed by a team of experts under the leadership of Prof. M. Surendra Prasad Babu of Andhra University in collaboration with N.G.R. Agriculture University. This website was launched in the internet on 3rd June, 2005 by Dr. Mangala Rai, D.G. of ICAR & Secretary of DARE, Government of India. The web portal is first of its kind containing the following features:

- This web portal provides time to time agricultural advice to the farmers at their door steps regarding selection of suitable crops and crop varieties for cropping, soil analysis, fertilizer management, pest and disease control, irrigation management etc., in their cropping pattern. It also advises the farmers for contingency crop planning to prevent the hazards and the organic farming system to provide residues free crop product.

- The agro met advisory system which is a part of the web portal will provide information to the farmers from time to time regarding the meteorological developments in the atmosphere well in advance to face a natural eventualities /calamities in an efficient way.
- The marketing advisory system which is also a part of the website will help the farmers to know the prices of their produces in and around them so that they can take their products to these places for their better market price.
- The web directory service and the discussion forum service provided in the website will help the scientific community in a greater way to interact each other to produce better findings in the agriculture field.
- Provides the information about the expert scientists, specialization wise, available in the different parts of the country and mailing facility to the farmers for contacting them for on line agricultural advice.
- Provides information about the various latest agricultural tools and implements, with their price & availability, so that farmers can buy them and use in their fields for the better productivity.
- The Website provides online computational facilities to the agricultural scientists to analyze their laboratory experiments (or on the fields)
- Provides facility to establish online journals for Indian agricultural scientists to publish their articles.
- Provides information about the Governmental schemes & facilities available to the farmers.
- Provides a rich database in the area of agriculture and help the country to lead the developing world.

At present the website www.indiakisan.net is good model and it is now for the industry, governments, national and international funding agencies, philanthropists and other scientist are to come together for making the dreams of Mahatma Gandhiji 's "Grama Swaraj", Smt. Indiraji's "Garibe Hatao" and Rajivi Gandhij 's Modern 21st Century world' into a reality.

ITC e-Choupal

International business division of ITC (Indian Tobacco Company) has conceived e-choupal, launched in June 2000 as a more efficient supply chain aimed at delivering value to its customers around the world on a sustainable basis. ITC's e-choupal is a unique example of using ICT's for agricultural development; e-choupal has already become the largest initiative among all internet based interventions in rural India. e-Choupal link rural farmers directly for the procurement of agricultural / aquaculture produce like soya, coffee, prawns etc. eliminating the role of the middleman. The principle of the e-choupals is to inform, empower and compete. There are 6,500 e-choupals today. ITC Limited is adding 7 new e-choupals a day and plans to scale up to 20,000 e-Choupals covering 100,000 villages in 15 states, servicing 15 million farmers.

Tara Haat

Tarahaat.com is a portal of Technology and Action for Rural Advancement (TARA), which is the marketing arm of the Delhi-based NGO Development Alternatives (DA). Tara Haat was launched in 1999. Tara markets information on products and services through Tarahaat.com. The content is developed by DA. Users can also set up their home page in Tarahaat.com. The aim is to create a virtual pipeline for delivery of not only DA information but also of other rural-based companies. The information can be accessed through telecentres called Tara kendras. There are 38 such Tara kendras in Punjab's Bathinda district and the Bundelkhand region (primarily near to Jhansi) of India.

Services Offered by Tara Kendras: Information is offered on a range of issues that are relevant for everybody in rural areas. Subjects covered include health, livelihood, governance, water, education and even entertainment. The information is static and on agriculture it is related to best practices. Tara haat also offers value added information services, for a higher fee. The information is dynamic and this includes information on market prices, weather, astrology and an e-mail facility of note is Tara haat's educational service, Taragyaan.

LISNET

Land Information System Network linking all institutions involved in land and water management for agricultural productivity and production systems, which has now evolved as "Agricultural Resources Information System" during the Tenth Plan is being implemented through NIC;

PPIN

Plant Protection Informatics Network (cibrc.nic.in) is a Step towards Bio Security and Ever Green Revolution in Indian Agriculture. Major Components of PPIN includes Computerized Registration of Pesticides (CROP); Plant Quarantine Information System (PQIS); Information System for Integrated Pest Management (IIPM) for CIPMCs; Pests & Disease Monitoring System (PDMIS) for States; Establishment of ICT infrastructure (LAN, WAN); Capacity Building and Intranet Portal.

Other useful web based services for farming communities are FERTNET: Fertilizers Informatics Network facilitating "Integrating Nutrient Management" at farm level; VISTARNET: Agricultural Extension Information System Network; APHNET: Animal Production and Health Informatics Network; FISHNET: Fisheries Informatics Network; AFPINET: Agricultural & Food Processing Industries Informatics Network; Weather NET: Weather Resource System.

Other Indian Agricultural Websites

1. **IFFCO Live (www.iffcolive.com):** IFFCO Live is an information sharing initiative by IFFCO. One can expect the latest news, activities and updates from all the parts of India, around the world & primarily about the most important industry, Agriculture.
2. **Agriculture Statistics website (www.indiaagristat.com):** It is a comprehensive source for Indian agriculture statistics which is regularly updated. It provides authentic statistical information on sectors like agricultural education, agricultural export, agriculture census, agriculture prices, agricultural insurance, animal husbandry, agricultural marketing, horticulture production, agricultural wages and all other relevant agricultural statistics of India. It also provides the agriculture related news. The statistics available on this web site is widely used by Indian agriculture research institutes, Indian agriculture sector, agricultural economists and Indian agriculture universities for Indian agricultural information, agriculture research and agricultural production data.
3. **ISAP website (www.isapindia.org):** Indian society of agribusiness professionals (ISAP) is a non- government, non-profit organisation. It is a network of agriculture and allied sector professionals in India and developing countries. Its vision is to instill economic security and stability among farming community particularly small and marginal farmers through holistic development of agriculture and rural sector. The site is all about the Indian society of agribusiness professional (ISAP) which

works for helping the farmers' community and address the rural - urban income divide.

4. **FCI website (www.fciweb.nic.in)**: This is the site by Food Corporation of India. It works for effective price support, food security, price stabilization and distribute food grains through public distribution system for the benefit of both farmers and consumers.

5. **Fertilizer India: (**www.fertindia.com**):** This website gives in detail about fertilizer statistics in India, fertilizer prices and fertilizer marketing in India.

6. **Farmers Scheme (**http://www.upagriculture.org): This agricultural site shows new schemes related to farmers, facilities given to them, latest techniques, new agri policies, quality control, useful machines, insects/ diseases of crop and many useful information's related to agriculture.

7. **India Gateway Agriculture (**http://www.indg.in/agriculture): This website aims to disseminate useful information about improved technology to the farming community and service providers in the rural areas. It aims to create a platform for different levels in the rural agricultural landscape - farmers, cooperatives and professional bodies, farm machinery vendors, fertilizer and chemical companies, insurance regulators and agronomists, consultants, and farm advisors.

8. **Agriculture Exchange (www.agiexchange.in):** Agri Exchange is a trade portal where farmers, suppliers and vendors of various agricultural products and services connect with potential buyers.

CHAPTER 8

ICTs for Agriculture Development in North East India

Suchiradipta Bhattacharjee, Rebekka Syiem and R. Saravanan
Central Agricultural University, Imphal

Like many other developing countries, agriculture is the primary occupation of the rural people of North East India. They depend on farming but because of the nature of practices followed, it hardly goes beyond subsistence and so agricultural extension is an important thing in their lives. But in this particular part of the country basically because of the topography which makes it difficult to reach the clusters of villages deep in the hills, forests and valleys, it becomes more and more important to use technologies that reach them like mobile phones, TV and radio. Agricultural information is not something that is needed 24X7 round the year but require on the critical times of the year like before sowing seasons advisories, forecasting diseases and pests when the crop is standing in the field and during harvesting. Therefore, a number of projects using Information and Communication Technologies (ICTs) are helping the farmers and extension personnel to reach each other in better and greater frequency and efficiency. This does not only increase economic output in terms of production but also in terms of money saved in travelling to the agriculture departments, efficient use of inputs, *etc*. Among the ICTs, mobile phones are having a major impact in agricultural extension because of their higher reach among the rural people. But inspite of all these positive effects of ICTs in agriculture, it does suffer from some chronic problems like lack of infrastructure, technical knowledge among users, proper study on actual impact of the ICTs on agriculture & livelihood and most importantly the sceptical nature of the users. The projects need to be well planned to be more useful to the target group and to increase their effectiveness and efficiency for the agrarian population.

North East India: An Overview

Undulating hills, picturesque valleys, diverse culture and an abundance of flora and fauna – that's North East India where lies 'Scotland of the East' Meghalaya, 'Land of the Rising Sun' Arunachal Pradesh, with Kanchenjunga at its borders Sikkim stands tall, mainstream Indian culture exists peacefully in the valleys of Tripura, world famous Martial Arts of Manipur, the homeland of Assam Tea and one horned rhinoceros in Assam, Mizoram with the sounds of 'Cheraw' echoing from its hills and the falcon capital of the world Nagaland. Area of the region is 262,230 sq. km and population is around 3.8 million. The region share more than 4500 km of international boundary with China, Mayanmar, Bangladesh and Bhutan. The Siliguri Corridor, with an average width of 21 to 40 kms. connects the NE region with rest of the country. Diverse climatic condition prevails in the region ranging from sub-tropical to temperate climatic regions. Mausynram in Meghalaya receives the highest rainfall in the world.

Situated in the midst of such a large reserve of flora and fauna, it is no surprise that 70 per cent of the population depends on agriculture for their livelihood (ICC, 2013). But it is mostly subsistance form of cultivation with Jhum or shifting cultivation. Agricultural development have been very low in the region due to unorganised cultivation of crops, inadequate infrastructure, low adoption of modern technlogy, lack of agricultural land and lack of knowledge. Inspite of the region's high dependency on agriculture sector, it still depends on other parts of the country for its own consumption of foodgrains. The farmers in the rural parts and interior hills rarely get the advice of the extension personnel mostly due to the difficulty in communication. The low density of population and hilly terrains makes it difficult to connect every locality. Many times it becomes impossible for the extension personnel to reach them because of climbing hills, rowing in the river for hours and travelling through forests. Even though the rugged terrain makes the human reach difficult. Technology has their presence in these parts. Radio, then TV and the latest additions are mobile phones and computer, allthough in very less numbers. Due to rising education, awareness has increased amongst the people resulting better livelihood. In this quest, the people of North East have been aided by several development agencies and Information and Communication Technologies (ICTs).

ICTs initiative in North East India

A number of ICT projects have been implemented in the region in the recent past to develop the agriculture and the farming community. Information is the most scarce soft resource for the farmers of the region and that makes their profession non-remunerative and puts livelihood in trouble. Looking the information need of the area various ICT projects have been implemented.

1. **e-Arik**: e-Arik, meaning e-Agriculture in Adi dialect of Arunachal Pradesh, was initiated in 2007 to disseminate 'climate smart agricultural practices' in the Siang river valley and foothills of Eastern Himalayas of North East India where few farmers have access to agricultural information, very few possessed mobile phones and none had computer or internet access. Under the project, Village Knowledge Centres were established in the Yangrung village with computer, internet link, printer, scanner, phone and TV. Project facilitators (agricultural professionals, a computer instructor and farmer facilitator) were appointed to help the farmers access ICT based agricultural information. In addition, the project portal (www.earik.in) also provided important information on crop cultivation and other agricultural practices; baseline information from relevant agriculture and rural developmental departments of government; specific information on government schemes such as farmer welfare programmes; and day-to-day market information and weather forecasts. Offline CDs also were available in the Village Knowledge Centres and the farmers had the facilities to engage in remote consultation with agricultural experts with the help of the infomediaries.

   The project was sponsored by Department of Scientific and Industrial Research (DSIR), Ministry of Science and Technology, Government of India. College of Horticulture and Forestry, Central Agricultural University was the project implementer with farmer facilitators, Farm Science Centre, Village Tribal Council members and 500 registered farmers from 12 remote tribal villages as stakeholders.

   The main impact of the project has been the adoption and implementation of climate smart practices by 44 per cent and 92 per cent of rice and mandarin farmers respectively. Two years after the project, 55 per cent of the farmers have developed new Khasi Mandarin orchards in their rice fields; 42 and 29 per cent have reported increased production in rice and mandarin respectively, and an average increase of annual income per farmer was reported to be US$ 37.50 and US$ 117 respectively in rice and mandarin respectively. Overall, e-Arik approach was found to be 3.6 times cheaper than the conventional extension system (Saravanan, 2011).

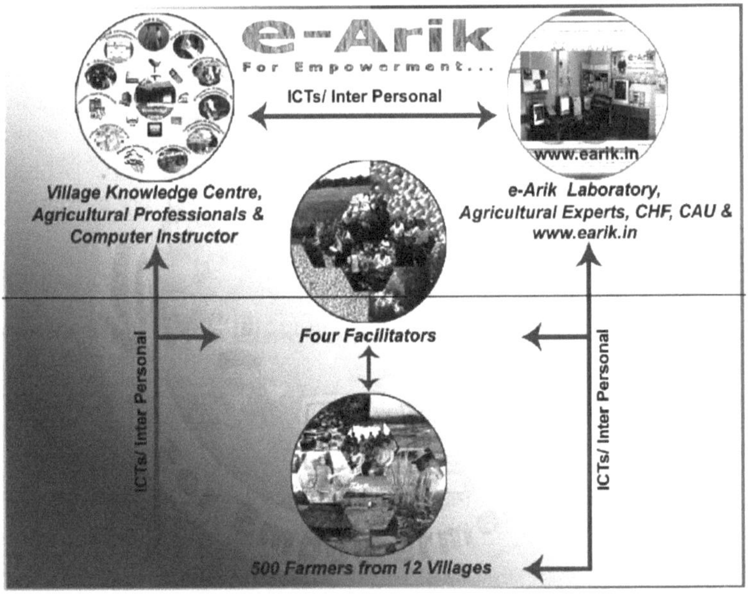

Fig. 1: Schematic diagram of e-Arik

2. **Model e-Village**: With the vision of escalating agricultural growth through generation of information on new technology & methods and to spread e-literacy among the disadvantaged tribal community, the pilot phase of e-village project was started in ten villages of East Siang District of Arunachal Pradesh state of India. The strategic location of the village which made it easier to access by the villagers from neighbouring villages was the reason for its selection. Model e-village project was jointly designed and implemented by Centre for Development of Advanced Computing (C-DAC) and College of Horticulture and Forestry, Central Agricultural University, Arunachal Pradesh and funded by Department of Electronics and Information Technology, Ministry of Commerce and Information technology, Government of India. The main objective of the project was to create e-awareness and e-literacy among the tribal community, creation of computer infrastructure, explore appropriateness of rural ICT application, and studies the use and impact of Indian Development Gateway's (InDG) products and services specific to North East India. The e-villages were equipped with computer, printer, scanner, wireless internet connection (EVDO) and DTH (Direct-to-Home) TV. A mini library was also established with technical publications in agriculture and allied sectors. The community took up the responsibility of providing location, furniture and electricity for the centre. Ten Project Facilitators and two Research Fellow were recruited by CHF, CAU to oversee the

functions of the e-Village centres whereas Project Investigator provided technical support in design and deployment and preparing e-learning modules. The project worked in close collaboration with farm Science Centres, NGOs and Government departments to bridge digital divide.

During its three years of operation, the project has provided farmers with both hard and soft resources. While the Information Technology (IT) centres in the remote villages gave the youth both employment opportunity and helped in computer skill building, the e-awareness programs conducted have benefitted more than 800 students. New entrepreneurial skills and technology in agriculture was introduced through multimedia CDs, leaflets, booklets, *etc*. Twenty eight per cent adoption of the new technologies among the farmers was observed and a 12 per cent increase in overall income level was indicated. More than 70 per cent of the school children aged between 9 - 17 years obtained education on basic computer applications (One World Foundation India, 2012).

3. **m4agriNEI**: It is a project of Media Lab Asia and College of Post-Graduate Studies (CPGS), Umiam of Central Agricultural University, Imphal to enhance the productivity and to empower the farmers by providing right information at the right time. The ICT initiative operates *via* a mobile based push and pull system where agriculture related information can be pulled/ pushed by the farmers using their mobile phones. The presence of a mobile interface at the front end of the farmers and web interface at the back end for the agricultural experts facilitates exchange of data through voice, text, images and videos from both ends. Farmers can also call the system through a toll free number to get any information and agro advisory. A centralised database is also available comprising of all the information of the farmers related to farm and previous transactions to enable experts at the back end to access to the database of the farmers while responding to their queries. The ICT Initiative also operates through a farmer coordinator present at the village level in order to help other farmers to get access the information through the use of ICTs. Further, the project staff of m4agriNEI regularly visited the farmers field to provide on-the spot solutions to the farmers related to farming. Presently as on January 2014, around 1191 calls have been received through the system, 1186 number of advisories are provided to the farmers and 1173 number of queries have been solved by m4agriNEI (www.m4agrinei.in, 2014).

4. **Community Information Centre (CIC)**: To provide the important facilities of Information and Communication Technologies to remote North Eastern region of India, Government of India has set up the

Community Information Centres in 2002 at block level in eight North eastern states of India. It is world's largest tele-centre initiative with a total number of 487 CICs established in the first phase. The project is aimed to set up ICT infrastructure at block level, to provide web access and internet services such as e-mail, market access and e-commerce facility and access to socio-economic databases. In addition, e-learning and e-education, e-medicine, e-consultation, e-governance, Government to Citizen (G2C) services and weather information are also available in the CIC centres. Each centre is equipped with computer communication equipment, a server and five clients connected to VSAT on a local area network (LAN), a TV and other ICT tools. Each CIC has two operators for managing and providing services to the public.

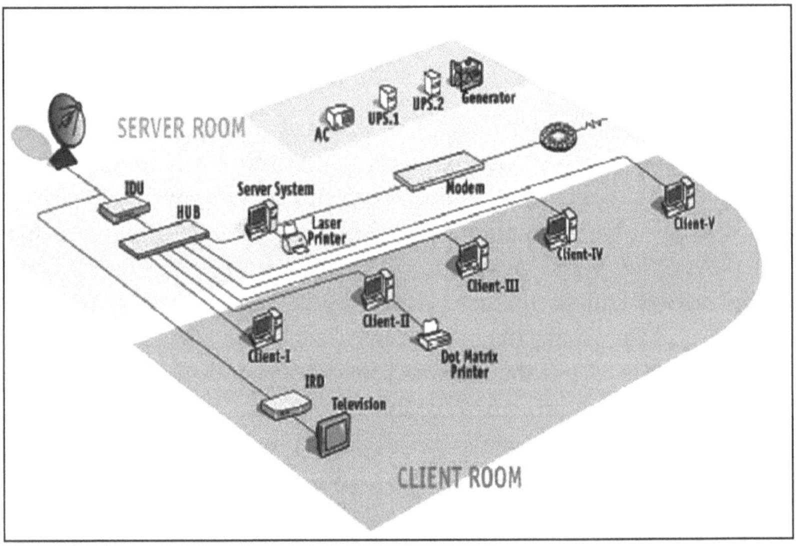

Fig. 2: Schematic representation of the working of CICs

The G2C service of the CICs is to take Government services to the citizen's doorstep through a one-stop service facilitation window called e-Suvidha. The web based application enables citizens to submit requests for Government services and monitor the status of their applications. In Assam, National Informatics Centre (NIC), Assam and Assam Small Farmers' Agribusiness Consortium has developed the model 'ASHA – Making Farmers Prosperous through CIC using ICTs' for facilitating agribusiness in the state to help the farmers and rural youth. While the project has won several accolades in national and international fronts, in NE India it has helped in rural development, agribusiness, medicine, e-governance and also has established the highest cyber cafe in the world in Sherathong, Nathula Pass, Sikkim (Chaudhri and Dash, Unknown). The present status of CICs in North East India is given in Table 1 below.

Table 1. Status of CICs in North East India

S. No.	State	No. of CICs
1	Assam	215
2	Arunachal Pradesh	85
3	Manipur	39
4	Meghalaya	39
5	Mizoram	36
6	Nagaland	52
7	Sikkim	45
8	Tripura	40
	Total	**551**

5. **Intelligent Advisory Service to Farmers (IASF)**: In order to harness the potential of mobile phones in the hilly terrains of North East India and make agricultural advisory mobile, IASF has been launched under the m-Gov initiative of Department of Electronics and Information Technology under the Ministry of Informtion Technology, Government of India. IASF is currently catering to the needs of 1886 farmers in six districts of Meghalaya and nine districts of Manipur who uses the platfom for redressal of their queries. Fifty Subject Matter Specialists are also registerd in the portal to respond to farmers' queries.

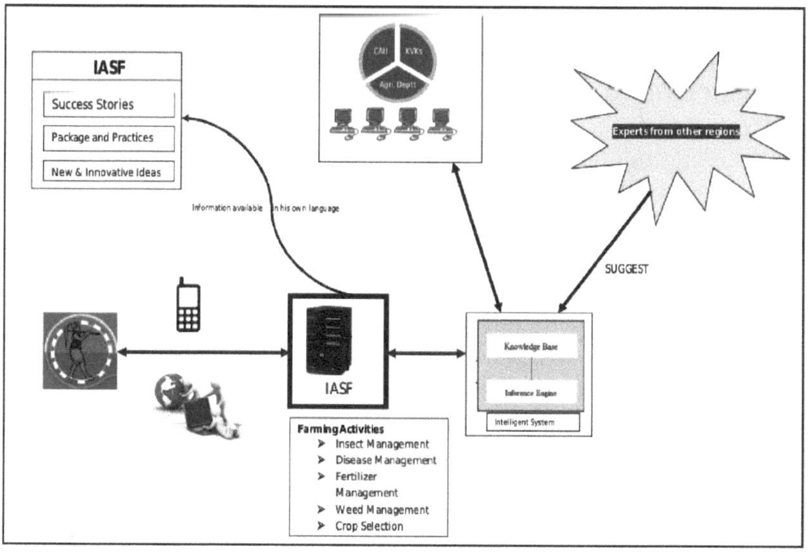

Fig. 3: Diagrammatic representation of functioning of IASF

The project is funded by Department of Information and Technology, Government of India and launched in collaboration with Centre for Advanced Computing (CDAC), Mumbai, Department of Agriculture, Government of Meghalaya, Krishi Vigyan Kendras of Meghalaya, Department of Agriculture, Government of Meghalaya and Central Agricultural University, Imphal. A special feature of the service is the 'Mobile Crop Doctor' in which a registered farmer can send a direct query related to any farming activity supported by IASF in the form Short Messaging Service (SMS) to 51969. All the registered crop experts will receive the SMS query from the farmer but only the first SMS answer sent by the registered crop expert will be forwarded to the farmer. The farmers' queries related to Insect Management, Disease Management, Weed Management, Rice Variety Selection and Fertilizer Management along with their solutions are stored in a database (CASE), thus forming a large database which the farmers can access anytime in case of contingencies (IASF, 2013).

6. **Kisan Mobile Advisory Service**: Kisan Mobile Advisory Service (KMAS) is an internet based SMS delivery system started by ICAR in 2007. KMAS is an attempt to integrate ICTs in extension delivery and to make it more efficient. This ICT based alternate agricultural information and rural delivery mechanism was initiated during 2007 in Zonal Project Directorate (ZPD) Zone VII, Jabalpur. The service is based on the organic model of communication which involves four major components of communication process. *i.e.* Sender, Message, Channel and Receiver. This is a unique programme for making linkages between several stakeholders who is main player for making Indian agriculture sustainable in the coming future through intensive use of ICT tools like mobile phones. Short Message Service is being provided by the KVK Subject Matter Specialist and when the institutions like ICAR are integrated, even the scientists also send the messages to increase the speed of diffusion of research findings.

   This is a unique program for making linkages between different stakeholders through intensive use of ICT tools like mobile phones using the feature of SMS for KVK-SMS/Scientists (Information Generator), Extension Functionary (Information User) and Farming community (Implementer). Keeping in view the importance of KMAS and the constraints in the farmers' field, the KMAS has been conceptualised with the following working objectives:

i) To measure the speed of rural information delivery system.

ii) To disseminate the need-based, timely information to the rural masses.

iii) To minimize information loss by making direct contact.

iv) To analyze the swift feedback mechanism of KMAS.

v) To assess the impact of the KMAS model.

A study conducted to understand the utility and feasibility of the service in rural Meghalaya found that due to lack of proper awareness initiatives from KVK, East Khasi Hills many farmers do not have a clue from whom they are receiving the messages and those who wanted to access or subscribe it did not know whom to contact. As of now, 65% of the farmers are not satisfied with the service but the scope to improve is huge in this area if timeliness, content of the message and personalization of the message is taken care of (Suchiradipta, 2012).

The following schematic diagram represents the flow of information in the KMAS model:

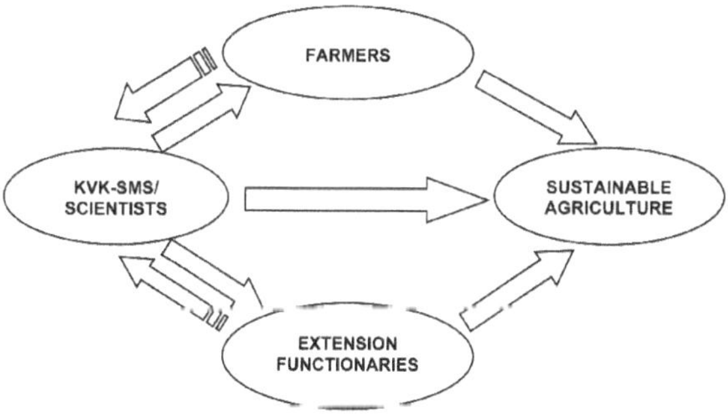

Fig. 4: Schematic diagram of KMAS model

7. **KIRAN**: KIRAN, the 'Knowledge Innovation Repository in Agriculture for North East' was launched in 2012 by the premier research institute of the country Indian Council of Agricultural Research in its Research Complex for North east Hill Region at Barapani, Meghalaya. KIRAN is basically a user platform and a warehouse repository for storing and retrieving information by agricultural stakeholders. Information relating to agriculture and allied sector in multi-media format is available in the KIRAN portal for anyone to access at any time. The platform will also help to foster linkages among public, private, state and regional organization functioning in North East India to help unleash the true potential of agriculture in the region (KIRAN, 2014).

Advanced Messaging System (AMS), as part of the KIRAN initiative was launched in 2013 for delivering advisory on agricultural and crop management practices and agro meteorological advisory to the farmers of North East India. So far, the messaging service has been initiated among 400 farmers from districts of Meghalaya on a pilot scale. It is envisaged to include a total of 2475 farmers all over the region. The messaging service would be contextual and would deliver season specific agro-husbandry practices, relevant to farmers in NEH region. Advisories regarding practices, problems and interventions would also be given on every alternate day to the tribal farmers through an SMS portal designed under the KIRAN platform (BioSpectrum, 2013).

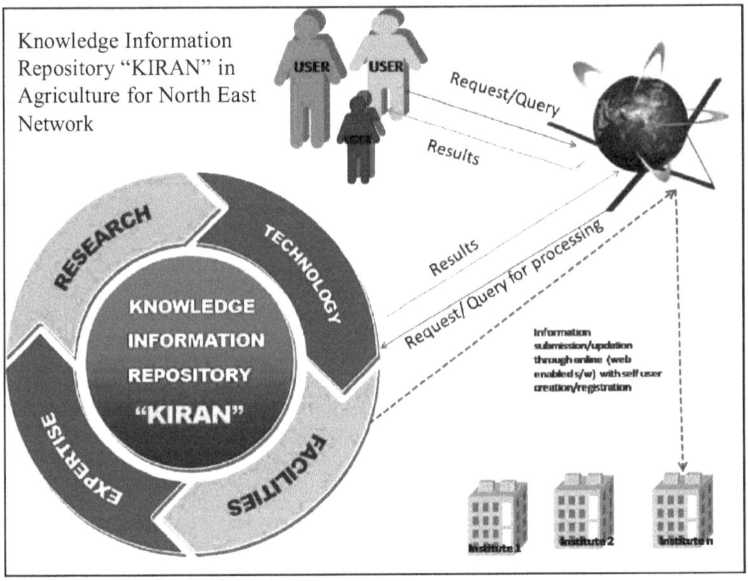

Fig. 5: Working structure of KIRAN

8. **Kisan Call Centre**: One of the earliest initiative in the country to integrate telecommunications in extension service delivery resulted in Kisan Call Centre in 2004. The purpose of this initiative of Department of Agriculture and Cooperation, Ministry of Agriculture, Government of India was to issue response to farmers in local language. Eight nodal centres were established around the country in Mumbai, Kanpur, Bangalore, Chennai, Hyderabad, Chandigarh, Delhi and Kolkata with Kolkata acting as the nodal centre for the eight North Eastern states till 2006. In 2006, it was shifted to Guwahati and Assam Agricultural University, Jorhat was made the nodal agency for looking after the functioning of all the KCC in the

eight states. The KCCs were established with the objective of bridging the gap between the extension-farmer linkages in efficient technology transfer and to help establish a network of relationship among the scientists, policy makers, extension workers, farmers and other stake holders to exchange their views to evolve a unique and revolutionary development of agriculture sector in the country.

The Kisan Call Centre, consists of three levels – namely **Level-I** (the basic Call Centre interface, with high quality bandwidth and local language proficient Agriculture Graduate sitting at KCC), **Level-II** (Subject Matter Specialists on concerned important crops and enterprises, connected through incoming telephone line) and **Level- III** (the Management Group to ensure ultimate answering and resolution of all the farmer's query which are not resolved at Level-II, connected on off line mode). The call centre is equipped to receive the farmer's query during all seven days. The incoming call is picked up by an operator (level –I expert) who after a short welcome message takes down the basic information and the query of the caller. These details are fed into a computer located next to the operator by the operator himself. The first level operators preferably are agricultural graduate with rural background knowing local language. They should also posses' good communication skills. They would be in a position to answer a majority of the questions likely to be asked by the farmers. Level–II consists of Subject Matter Specialists (SMS) who are located at their Institutes/Departments identified as Response Centres and arranged to be available regularly in all working days during working hours on roster basis to provide the necessary reply. In case level-I personnel is not able to answer the question, he forwards (in call sharing mode) the call to the concerned Subject Matter Specialist. The question asked is transferred to the Level-II functionary to reply the query. In case, it is not possible to answer, there is a system to revert to the caller by post / fax / e-mail or by telephone. Level–III consists of a dedicated cell located at the Nodal Office to receive the questions that have not been answered at the first and the second levels. Appropriate replies to these questions would be then framed in consultation with the concerned specialists available within or outside the State, by the nodal cell. The replies would be sent to the farmers promptly by post/e-mail/ fax/ telephone *etc*. (DAC, 2014; www.manage.gov.in, 2014).

Farmers Dial Helpline

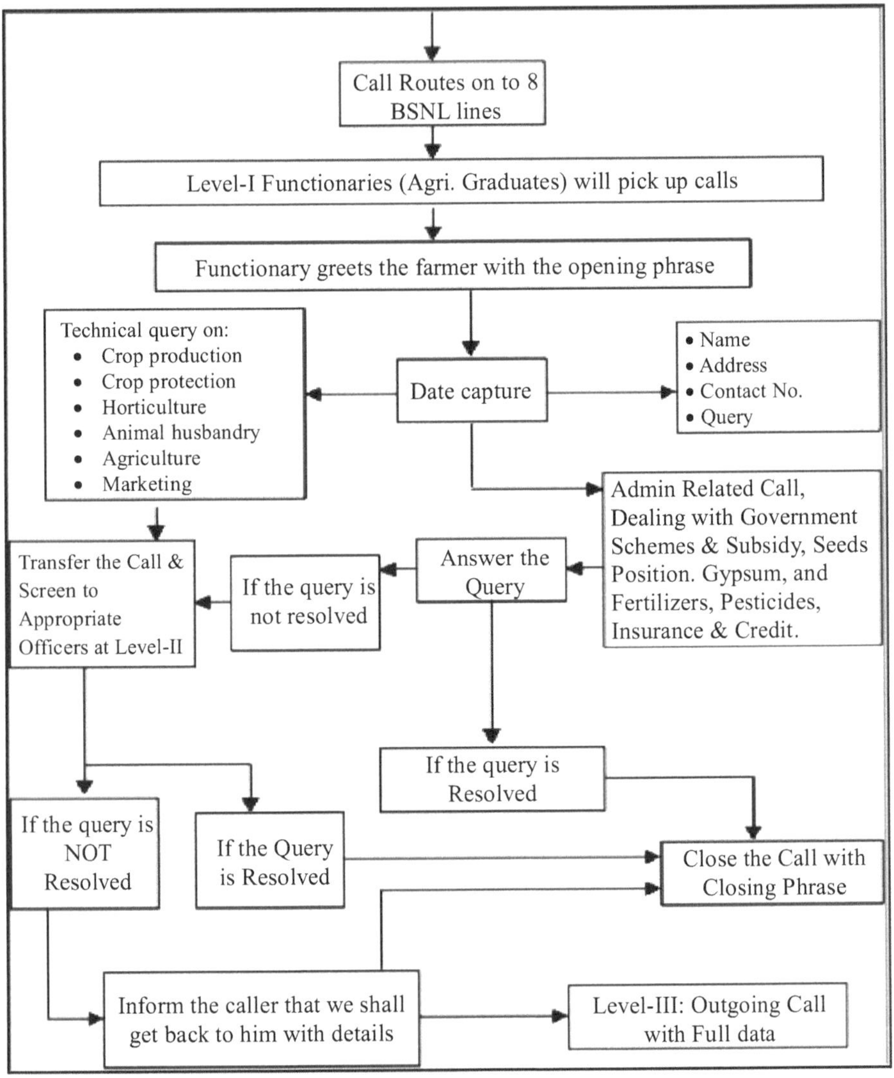

Fig. 6: Schematic Representation of Kisan Call Centre Operational Mechanism

Conclusion

Information and Communication Technology (ICTs) have helped agriculture in North East India change for good but only in some parts. Replication of success has been very less. But inspite some drawbacks, it has been able to bring the change in the old and the young. But ultimately technology goes as far as we take it and it is upon the shoulders of all involved to take it forward for mainstreaming agriculture and communication technology in North East India.

References

Bio Spectrum, 2013. KIRAN launches Agricultural Messaging Service from ICAR Barapani. Retrieved from http://www.biospectrumindia.com/biospecindia/news/189830/ kiran-launches-agricultural-messaging-service-icar-arapani#sthash.sm0cMfGL.dpuf (Updated on 11th June, 2013; Accessed on 13th January, 2014).

Chaudhri, N. and Dash, S. S. Unknown. Community Information Centres. Retrieved from http://www.csi-sigegov.org/casestudies/13_cics.pdf (Accessed on 13th January, 2014).

DAC (Department of Agriculture and Cooperation), 2014. Kisan Call Centre. Retrieved from http://djd.dacnet.nic.in/kccdetail.html (Accessed on 13th January, 2014).

IASF. 2014. Intelligent Advisory System for Farmers. Retrieved from http://iasf.cdacmumbai.in/ias/index.jsp (Accessed on 13th January, 2014).

ICC (Indian Chamber of Commerce). 2013. India's North East: Diversifying growth opportunities. Retrieved from http://www.pwc.in/en_IN/in/assets/pdfs/ publications/2013/north-east_summit-2013.pdf (Accessed on 10th January, 2014).

KIRAN. 2013. KIRAN – Empowering agricultural innovation and knowledge in North East. Retrieved from http://www.kiran.nic.in/about_us.html (Accessed on 13th January, 2013).

One World Foundation India. 2013. Model e-Villages in North East India: An ICT project for development in remote tribal areas. Retrieved from http://indiagovernance.gov.in/files/e-model-village_gkc.pdf (Accessed on 14th January, 2014).

Saravanan, R. 2011. e-Arik: Using ICTs to Facilitate "Climate-Smart Agriculture" among Tribal Farmers of North East India.

Suchiradipta, B. 2012. Mobiles for mobilizing agricultural extension in India. M.Sc. Seminar presented at School of Social Sciences, College of Post Graduate Studies, CAU, Umiam, Meghalaya.

Wikipedia. 2014. North East India. Retrieved from http://en.wikipedia.org/ wiki/North_East_India (Updated on 3rd January, 2014; Accessed on 10th January, 2014).

www.m4agrinei.in. 2014. m4agriNEI for better farming. Retrieved from http://www.m4agrinei.in/ . (Accessed on 16th May, 2014)

www.manage.gov.in. 2014. Kisan Call Centre. Retrieved from http://www.manage.gov.in/kisan/Operational_mechanism.htm (Accessed on 13th January, 2014).

CHAPTER 9

Mobile Phone as ICT Tool

A mobile phone often referred to as a cellular phone or cell phone is an electronic telecommunication device that can make and receive telephone calls through radio wave or satellite transmission whilst moving around a wide geographic area. Most mobile phones provide voice communications, Short Message Service (SMS), Multimedia Message Service (MMS), short range wireless communication (Infrared, Bluetooth), business applications, gaming, photography and internet services such as web browsing and e-mail. Mobile communication technology has become the world's most common way of transmitting voice, data and services, and no other technology has ever spread so fast. The mobile phone is especially important for India because it is growing here faster.

The cellular technology is becoming popular day by day due to increasing number of companies with low price handsets and decreasing rate of tariff plans. Nearly one fourth of the country's population possessed mobile phone even in the rural area. According to press release by Telecom Regulatory Authority of India (TRAI) total wireless subscriber base had increased to 934.09 million at the end of June 2012. The share of urban wireless subscribers was 63.98 per cent whereas; share of rural wireless subscribers was 36.02 per cent. The overall wireless teledensity (number of telephone subscribers per 100 individuals) in India has reached 76.99 in June 2012.

Present scenario of mobile application in dissemination of agriculture information indicates that mobile phone will go a long way in enabling the farmers to receive timely information. Currently, cell phones are widely used in day to day activities for social, economical and political spheres of life. Keeping in view enormous scope of mobile phone application in agriculture extension and wide penetration of mobile phone among the farming communities, an effort was made to explore the present status of application of mobile phone to agriculture by various stakeholders in India.

Why Mobile Phone for Agriculture Extension Service

Our recent research reveals that the mobile phone is playing a very useful role in fulfilling the informational needs of farmers, particularly among marginal and small ones. The mobile-phone based agricultural information services are now becoming popular among the farming community. These services, through SMS or voice-messages provide a variety of agriculture related information on crop-cultivation, fertilizer use, plant-diseases, pesticides, market-prices, weather and important government policy decisions. The information is provided to farmers in local language, within a specified time and also two-way interaction through customer care centers is available. The farmers who have subscribed to these services have highlighted that they have now been more aware and have also enhanced agricultural earnings. The farmers who are not the subscribers but possess a mobile phone also revealed that the instrument has helped reduce costs and wastages and increased incomes. The popular uses of mobile phone in agricultural operations (when used just as a communication medium) included, getting to know the market-prices of crops at various places; receiving instantaneous solutions regarding seed-variety, fertilizer and pesticide availability/application; calling distant livestock-doctors and so on. Significant saving in both time and money/fuel were reported by farmers on account of mobile communication. The research also provides evidence on the key role that mobile phones are playing in improving the information transfer between farmers and research institutions, government & private input companies, input-dealers, doctors, markets and other farmers. The mobile based agricultural services are providing enthusiastic response from the farming community and farmers have willingness to pay for these services. The mobile phones technologies have huge potential to revitalize our agricultural extension services. Mobile phone is becoming a very popular mean for information exchange in rural area therefore; both public and private organizations have started to use mobile phone as channel for transfer of agricultural technologies among the farming communities. Major mobile phone based initiatives taken in India are discussed here. Review of these initiatives would provide more ideas of mobile phone application in agriculture.

1. **IFFCO Kisan Sanchar Limited (IKSL):** The Indian Farmers Fertilizer Cooperative Limited (IFFCO) launched a project to leverage information and communication technology, in particular mobile telephony to disseminate expert agricultural knowledge to small-scale farmers in rural India. The newly created IFFCO Kisan Sanchar Limited (IKSL) is a

joint venture between IFFCO and Airtel. In India, farmers in rural areas have little access of knowledge about improved farming practices and rural development services. IKSL manages the distribution of mobile phones, SIM cards, and information services to small farmers and uses of a sim card with a voice message platform to record content and send information to farmers through voicemail. Farmers get access five daily messages in their local language with crop and area specific information in relation to agricultural market prices and arrivals, availability of fertilizers, electricity timings, early warning systems on disasters, weather forecast, best farming practices, local crops management, education and upcoming training opportunities, plant and veterinary disease prevention, financing, insurance services and government schemes; a help line and interactive information services, all free of cost is also available. Voice messages in local languages 95,000 voice messages delivered and 81000 Q & A repository with 5000 feedback messages from the farmers. 10 Lakh active farmers benefiting from IKSL's Value Added Services and IKSL enrollment crosses 4 million and 40000 cooperative societies as IKSL retailers (Source: www.iksl.in). For more details about IKSL see chapter no. 07.

2. **Fisher Friend Project:** The Fisher Friend project of the M.S. Swaminathan Research Foundation (MSSRF) in Tamil Nadu and Puducherry leverages mobile technology to provide vital livelihood information to fisher folk. MSSRF partnered with Qualcomm, Tata Teleservices and Astute Technology Systems for developing the Fisher Friend Mobile Application. The tool was designed after a thorough needs assessment of the fisher communities and incorporation of feedback from central stakeholders. Upon sending a single button click request from an icon based software module on mobile, fishermen gain access to vital updates on wave height, wind speed and direction, potential fishing zones, news, government schemes and market prices. All content is displayed in the local language Tamil. This unique application is helping fisher folk make better choices and avoid hazardous situations. It is enabling them to conduct their livelihood operations in a safe and profitable manner. MSSRF is also working on setting up village knowledge centres from 1992. A needs assessment study revealed that locale-specific, dynamic content such as market prices of fish or best fishing zones were in great demand. While the VKC had been able to provide this information in the village itself, fishermen away at sea do not have access to it. It was during discussions on how to expand the dissemination of such information that the idea of using mobile phone technology cropped up. On the third anniversary of the tsunami the Mobile Fisher Friend project

was started with the Foundation handing over 40 mobiles to fisher folk of Puducherry, Chidambaram, Nagapattinam, Manamelkudi, Thangachimadam, Vembar and Nagercoil.

3. **RML' SMS Service:** Reuters Market Light (RML) an initiative started by Thomson Reuters, is an information service started in 2007. RML targets information provision to Indian farmers via mobile phones. RML subscribers receive information via text messages on mobile phones, which is personalized based on the stage of the crop cycle and at the taluka/ block level for weather forecasts. They also provide information about local market prices and local and international commodity information. Since personalization is a key factor in this service, one prominent feature of RML is that this information can be accessed by the farmer in his local language. So overall, a farmer gets much of the necessary information in his language through mobile phone SMS.

   This initiative has recorded one million unique subscribers so far in an estimated 50,000 villages across 13 states. RML believes that farmers tend to prefer text messages over a phone call as that information remains stored in the phone and can be understood more easily. But the only bottleneck for this interface in a country like ours is that of literacy. This limits the range of people who can gain benefits from such a scheme. This initiative is an example of how making the right information available to the farmer in a timely fashion can help them improve farm yields and hence improve their conditions. A study conducted by Indian Council for Research on International Economic Relations (ICRIER) claims that they observed 5-25% increase in the incomes of the RML subscribers interviewed by them.

   The broad services provided to farmers include, localised - commodity pricing (Onion, Cotton, Soybean, Pulses, Pomegranate etc.); weather updates; news (agriculture & general) and service is available on GSM networks only, but not on CDMA networks. Reuters Market Light is a pioneering mobile phone-based highly- personalised professional information service specially designed to support the farmer community.

   RML was launched in India in the state of Maharashtra in October 2007 and in October 2008 in the state of Punjab. Today RML operates in 13 states in India and covers over 250 crops and more than 1000 markets and 2800 weather locations across these states.

4. **AGMARKNET's E-Alert:** AGMARKNET is a Central Sector Scheme to network Agricultural Produce Wholesale Markets (APWMS), State Agricultural Marketing Boards/Directorates and DMI Regional Offices to collect agricultural produce price/arrival information and dissemination. AGMARKNET portal is accessible to all stakeholders through internet access. Recently AGMARKNET have started e-Alert service which is a method to facilitate and access agricultural produce price of market information by the public through e-mail or SMS. The subscription of the service is free for all. To subscribe for e-Alert on market prices of agriculture produces one have to get registered as new user by using e-Alert login page on URL: http://agmarknet.nic.in. At the time of registration the users have to select type of alert he/she wants (through e-mail, SMS and both), Days of week on which you want the alert, type the period up to which you want alert, You can opt up to 6 different combinations of markets, commodities and varieties for one mobile phone number. Chosen combination, can be changed user by using SMS registration link page.

5. **TCS' mKRISHI**: mKrishi is a mobile based advisory platform. It was started by Tata Consultancy Services initially as pilot projects in Maharshtra in October 2006. mKrishi initiative uses mobile phones and the sensor technology to give personalised advice to farmers. It was felt that mKrishi had the potential to create new markets and offer its services at a low cost therefore; TCS launched mKrishi in 2009 in the agricultural states of Punjab and Uttar Pradesh, where it now serves thousands of farmers. mKrishi platform combines multiple technologies to bring vital information regarding local weather, fertilizer requirements based on soil conditions, pest control, and current food grain prices in local markets in a rich content format to the farmer's low-end mobile handsets. It allows farmers to send queries in their local languages, as well as images and voice activated SMS through a mobile phone and provides personal responses with advice or relevant information in these languages. It was realized that the farmers required several services besides agro advisory, such as poultry, cattle, micro credit, crop insurance, government policies, storage and selling of crops. mKrishi system was conceived to address all these challenges. It is a paid service costs a farmer between Rs 40 to 80 per month. TCS partners with wireless operators allow farmers to download the platform on high-end phones, and TCS has set up "mini-mobile sites" that farmers can visit to have the platform installed on low-end phones. The mobile phone overcomes the lack of power and wire-line communication infrastructure in rural areas, enabling farmers to get advice one-on-one from experts.

The platform's technology not only allows farmers to submit questions to experts, but also provides environment-specific details that give the experts a kind of agricultural map of the issue at hand. For example, when a farmer enters his location on mKrishi, agricultural sensors connect to geo-location services like GPS and Google Earth to deliver local weather, soil conditions, common pests and food-grain prices to the expert on the other end. Farmers can also attach photos if they have mobile cameras. TCS' mKrishi platform uses mobile technology to cater to the absolute needs of the rural sector.

Thus, mKrishi initiative offers personalised and integrated services in local languages to farmers on their mobile phones. The service aims to enable farmers in remote areas to connect to their stakeholders, access good quality agricultural inputs, find advice on farming practices, and get information on market prices, weather and other information that can help in planning and boosting yields.

How does mKrishi work

mKrishi is available in three versions: Lite, Regular and Plus.

mKrishi Lite: Lite is an entry level service, with an Interactive Voice Response (IVR) based system. The farmers dial a published number to avail this service. They listen to and select the appropriate audio prompts in the local language of the region. They record their questions and get a question ID for each question. Experts analyze the queries and provide the appropriate audio advice. An SMS is sent to the each farmer indicating that his / her question has been replied to. When the advice is available, the farmers enter the query ID on the phone to listen to the expert's advice. Experts can also send out various alerts like the weather forecast, market prices, a package of practices, etc. in either English or the local languages to the farmers.

mKrishi Regular: Regular is a mobile based service, wherein a mobile application resides on the farmers' handsets. All the farmer interactions for the service happen through this application, in the local language. Farmers can download the mKrishi application on their mKrishi compatible and GPRS activated handset. Upon starting the application, they have to provide some basic information about themselves and their farms. The menus are programmed in the local language.

mKrishi Plus: It offers Automatic Weather Station (AWS) within a farmer's environment, along with the mKrishi Regular service. The AWS sends updates about various weather parameters like rainfall, temperature, moisture, etc. related to the installed location, at regular intervals. This integration provides a more personalized environment for farms and is recommended for orchard crops such as grape and some citrus fruits.

Objectives of mKrishi

- Reach farmers individually to understand their needs.
- Enable farmers to receive important information about pesticides, fertilizers and soil and water conservation.
- Provide critical micro climate, weather information in order to plan farming operations.
- Facilitate better production and cultivation practices.
- Improve access to markets, refining the associated economy.
- Empowering the rural community.

6. **vKVK:** In vKVK or 'virtual Krishi Vigyan Kendra', tailor made information, relevant to the farmer are transmitted to them on their mobile phones through texts and voice messages in local languages. This information is provided to the farmers by the scientists from the KVK, to which they belong, providing assurance regarding the authenticity and applicability of the advisory- the expert to farmer (E2F) service. It takes place either in web to mobile (W to M) or mobile to mobile (M to M) form. Similarly farmer to expert (F2E) and expert to expert (E2E) services are also being launched.

7. **aAqua Mini** – It offers real-time decision-support tools to farmers and organizations supporting progressive farming. The project envisaged working on revenue generating business model. The services provided to farmers include, localized remote crop diagnostic solution; audio prompted guide application (in English/Marathi/Hindi); remote crop & land properties based disease diagnostics; weather information (temp, cloud cover, precipitation); SMS enabled register and query mechanism; online poll for registered users; spam, search, rank features; and service is available on GSM and CDMA networks. The aAQUA Service is a problem-solving system dedicated to find solutions to problems posed by Indian farmers. Answers to agri-related queries are sent in 24 to 72 hours depending on the difficulty.

8. **E-Mausam SMS Service:** Government of Haryana has launched 'e-mausam' website to provide weather forecasts and agro advisories using Web and SMS-based alerts through National Informatics Centre (NIC)'s SMS gateway. It is helping farmers to have more lead time to minimize losses due to abnormal weather conditions. Also this helps farmers to increase their productivity with efficient management of day-to-day farm operations. To avail the advantage of this, one has to simply visit http://

emausamhau.gov.*in* and get registered to receive SMS with weather condition details. The state government has also kept other options of registration in Krishi Vigyan Kendras (KVKs) and local Agriculture Officers by submitting the required details. The web portal was conceptualized by Ch. Charan Singh Haryana Agricultural University (CCSHAU), Hisar, with technical support from the National Informatics Centre (NIC), Haryana. So far, more than 75000 farmers of Haryana have been registered with this portal and availing SMS on weather forecasts.

9. **Kisaan SMS Portal:** During celebrations 85th ICAR Foundation Day, Hon'ble President of India launched on 16.07.2013 a SMS Portal for Farmers created by the Department of Agriculture & Cooperation, Government of India. Officers, Scientists and Experts of Department of Agriculture & Cooperation, Indian Council of Agriculture Research, Department of Animal Husbandry, Dairying & Fisheries, India Meteorological Department and Food grain Procurement Agencies will use this portal for disseminating relevant information, giving topical & seasonal advisories and providing services through SMSs to farmers in language of the State. The State Governments and their field formations down to the Block level (including State Agriculture Universities and KVKs) also have come forward in a big way to use this integrated Portal. Considering availability of more than 38 crore mobile connections in the rural area, Kisaan SMS Portal is likely to be very useful to farmers and all other stake-holders as timely and relevant information will be provided to farmers in their own language. The SMSs will get transmitted only to the farmers within the territorial jurisdiction of an officer, scientist or experts for the crops or agricultural practice that such a farmer might have opted for. Search facility (by text and by topic) for going through previous advisories has also been created to avoid duplication and contradiction in advisories. Fulfilling the information and technology needs of the12 crore farm families has always been a challenge to the Extension system in the country.

10. **Kissan Kerala:** It is an integrated, multi-modal Agricultural information system, which provides several dynamic and useful information and advisory services for the farming community across the state of Kerala. The core deliverable and achievements of the project is an integrated multi-component, multi-modal delivery of Agriculture Information Services System that is accessible anywhere anytime by all concerned. The project adopted a strategy of providing right information to the right people in the right context and empowers the farmers with adequate

knowledge, which helps them to take better decision. The project solves the problem of content gaps by providing the authentic agricultural information though various delivery methods like Television, Internet, Telephone, and Mobile. The farmers may choose any medium to seek the relevant information. The project offers many services through the effective integration of ICT systems and tools to reaching out to the farming community.

11. **Nokia Life:** It is having a range of agriculture, education and entertainment services designed especially for the consumers in small towns and rural areas of the emerging markets (Nokia report: 2008). This project works on private partnership (PP) based revenue generating business model in India. The services include information on seeds, fertilizers, pesticides, weather (temperature, rainfall, wind conditions) and prices in English, Marathi and Hindi language option and prevailing market prices, education service in dual language display option.

12. **Behtar Zindagi Service:** Bharti Airtel in association with Value Added Services (VAS) provider Handygo Technologies launched 'Behtar Zindagi' – a mobile based service for farming and allied communities across Rajasthan. Customers can avail this service by dialing the toll free number 556780 and avail the subscription pack at a charge of Rs 30 per 30 days. This service supports a wide range of dialects and is available in Hindi and 17 other regional languages. The 'Behtar Zindagi' service is available for farmers 24×7 and 365 days of the year. Behtar Zindagi offers services such as weather updates, mandi rates, live stocks, agriculture, fisheries, health, education and finance. The 'Behtar Zindagi' service will leverage the simple and easy-to-use Interactive Voice Response (IVR) medium to communicate key information to subscribers, thus allowing them to leverage their mobile phones to plan their businesses and lead better lives. *'Behtar Zindagi'* has been developed after a detailed research around the farming community and keeps in mind the information gap that currently exists in this segment. To launch this initiative, Airtel joined hands with Handygo to bridge the gap between static and dynamic information. To ensure A-grade quality information, Handygo has worked with various private and government bodies (including Indian Meteorological Dept, Indian National Centre for Ocean Information Services, CARE, EKO and NETFISH) to collate relevant information and organize its dissemination to farmers using region interactive voice as medium supported by voice and text messages.

13. **Live Mandi APPs:** India is called country of villages, In Indian village there will be no light at home but there will be a Mobile Phone. Still villages lack somewhere in terms of technological usage for various reasons. Farmers do not know at what rate their commodities are being sold. Live mandi APPs was developed to fill the information gap. With the help of this application farmers can know at what price or rate their commodities are being sold at their neighborhood. Open Government Platform (OGPL) has done excellent job to provide this live data. This application collects all the available data and display to end user, in user friendly manner. Application user interface has mix of English and Hindi language, because in our villages not all people know Hindi, and not all know English.

Our country needs more initiatives like these and we need to brainstorm on how this information can be communicated to every farmer of the country and give him the right guidance needed by him. It is implementation of simple ideas that can bring real change. RML realized the importance of local language. Provision of information in local language breaks the fear from technology also. Computers and mobile phones were not considered comfortable for rural India before Hindi fonts came into play. With language support, much more interest can be observed from user end. ICTs can definitely bring a significant change in Indian agriculture system if we use them smartly and understand the needs properly. RML's success story can be seen in a timeline on their site.

CHAPTER 10

Touch Screens and Information Kiosks: ICT Tool for Agriculture Extension Services

Touch Screen Kiosks

A touch screen is a device, which allows users to control a Personal Computer by simply touching the display screen by finger or pin. This type of input is suitable for a large number of computing applications. Most PC systems use a touch screen as easily as other input devices such as track balls or touch pads. If a touch screen computer system is installed at the kiosk it is known as touch screen kiosk. It may be Just like ATMs or modern plasma TV set which are developed as information kiosk and the information are made available to the users. A visitor can see the visuals and interact with kiosk to get the desired information. He can elicit the expert information by pressing keys or touching icons. Users can also follow the method of application of any technology through seeing of clippings at touch screen kiosk. In the touch screen kiosks, information is made accessible in the form of images, voice, video, graphics and texts. A large number of touch screen kiosks have been established by Indian railway, government offices, Banks and private companies where by touch of finger users can get access to various services (G2C and B2C) offered by the organization or company.

Touch Screen Technology

A touch screen is a computer display screen and also an input device. The screens are sensitive to pressure; a user interacts with the computer by touching pictures, Icons or words on the screen. It is a type of display screen that has a touch sensitive transparent panel covering the screen. Instead of using a pointing device like mouse or light pen, you can use your finger to point directly to objects on the screen. Presently touch screen technology is being widely used in the computers, laptops, tablets, smart phones, plasma TV, ATMs, information kiosks etc. therefore, primary knowledge of touch screen technology is the need of hour.

Types of Touch Screens: Commonly use touch screen technology in the information kiosks can be categorised in the following type of touch screens.

1. **Resistive Touch Screens:** Resistive screens are pressure sensitive and can be touched with a finger, stylus or just about any pointed object. They use two active layers: a flexible plastic layer on top of a rigid plastic or glass layer, with insulated spacers in between. The layers are coated with indium tin oxide, and different voltages are applied across the coatings, typically alternating between the layers. When touched, the front layer picks up the voltage from the back, and the back layer picks up the voltage from the front, enabling the controller to determine the X-Y location. Although the least expensive, the resistive method blocks up to 30 percent of the light from the CRT or LCD screen due to the multiple layers and coatings.

2. **Capacitive Touch Screens:** The capacitive method uses only one active layer: a metallic coated glass panel, thus allowing more light to come through. Voltage is applied to the corners of the screen, and when a finger touches the screen, it draws a tiny amount of current. The controller computes the X-Y location from the change in capacitance caused by that touch point. Because the human body absorbs current, the finger must be used and not a plastic stylus. A variation called "projected capacitive" uses a sensor grid sandwiched between two glass layers. When the screen is touched, the controller computes the X-Y location from the change in capacitance in the grid. The grid also enables two-finger touching like Apple's multi touch iPhone. Although the grid is embedded and protected, the screen can be overlaid with a clear, heavy-duty glass layer for more protection.

3. **Surface wave Touch Screens:** Surface wave technology uses ultrasonic waves that pass over the touch screen panel. When the panel is touched, a portion of the wave is absorbed. This change in the ultrasonic waves registers the position of the touch event and sends this information to the controller for processing. Surface wave touch screen panels are the most advanced of the three types, but they can be damaged by outside elements.

4. **Acoustic Waves and Infrared Screens:** Acoustic waves or infrared signals are transmitted across the screen's external surface from the top and side. When the screen is touched, receivers at the opposite ends form an X-Y matrix and sense the blocked signals. Since these methods do not use active layers over the screen, they do not block any light. The screens can also be overlaid with a clear, heavy-duty glass layer for protection.

How does Touch Screen Work: Touch screen has three main components: a touch sensor, a controller and a software driver. To make a complete touch input system, a touch screen is combined with a display and a PC or other device.

1. **Touch Screen Censor:** A touch screen sensor is a glass plate having touch responsive surface. The touch sensor is positioned over a display screen so that the responsive area of the plate covers the maximum viewable area of the video screen. There are number of touch sensor technologies available in the market today, each using a different approach to detect touch input. The sensor has an electrical current or signal passing through it and touching the screen causes change in voltage or signal. This voltage or signal change is used to find out the location of the touch to the screen.

2. **Controller:** The controller used in touch screen is a small PC card that interconnects between the touch sensor and the PC. Controller takes data from the touch sensor and converts it into information that PC can understand. For integrated monitors, the controller is usually installed inside the monitor or it is placed in a plastic case for external touch add-ons/overlays. The controller is useful to determine what type of interface/ connection you will need on the PC. Integrated touch monitors are provided with an extra cable connection on the back for the touch screen. Controllers are available by connecting to a Serial/COM port (PC) or to a USB port (PC or Macintosh).

3. **Software Driver:** The driver is software for the PC system that permits the touch screen and computer to work together. It tells operating system of computer how to interpret the touch event information that is sent from the controller. Today's touch screen drivers are a mouse-emulation type driver. This makes touching the screen the same as clicking your mouse at the same location on the screen. This permits the touch screen to work with existing software and provide new applications to be developed without the need for touch screen specific programming. Some devices such as thin client terminals, DVD players and specialized

computer systems either do not use software drivers or they have their own in -built touch screen driver.

Information Kiosks

Kiosk is a small enclosed structure, often freestanding, open on one side or with a window used as a booth to access to information in agriculture and allied areas. In generalized term information kiosks are the public installations wherein computers are installed to make G2C, B2C services accessible in general to the citizens and agriculture extension services in particular to the farming community. Information kiosk is a hub of information as per the need of the area or the best source of information. Any visitor at the information kiosks can have access to any kind of agriculture information regarding package of practices, plant protection measures, nutritional deficiencies and their management, symptoms of various pests and diseases of several crops and their control measures; any other problems he encounters in the field. Even video clippings on various agriculture technologies along with voice can also be viewed and information about technological applications can be listened in local language. A large number of information kiosks have been established in India by public and private sectors for making available various services under a single roof at their door steps.

Agriculture Extension Services through Information Kiosks

1. **Community Information Centres (CICs):** IT revolution is based on knowledge and knowledge is the only thing that can be shared without dividing. By sharing, it is not divided on contrary it gets multiplied. The CIC Project has been implemented on pilot basis in 30 blocks of the North Eastern states and Sikkim. The pilot project was inaugurated in August 2000. The main project in North Eastern states was inaugurated in 17th August 2002 with the dedication of 457 CICs to the people of the region. In view of the successful implementation of the CIC Project in the North Eastern States it was decided to extend the same to other parts of the country. The first state to be taken up was the state of Jammu and Kashmir. This was due to the fact that the barriers for access to ICT, such as difficult geographical terrain and lack of infrastructure, which were found in the North Eastern states, were also an issue for Jammu and Kashmir. It was felt that the basic needs of citizens such as information, education, entertainment and health services could be addressed through the establishment of Community Information Centres in the state. These centres would serve as a platform for e-governance, e-learning and other IT enabled services in the state. Each CIC has two

CIC Operators (CICOs) for managing the centers and providing services to the public. CIC operators have been trained to manage the centres, provide services and impart training on basic computer operations and software packages to visitors. The project is a joint effort by Department of Information Technology (DIT) under Ministry of Communications and Information Technology (MCIT), National Informatics Centre (NIC) and implementing State Governments of the North-Eastern states, J&K and AP. (*Source*: www.cic.nic.in).

Progress of CICs: Department of Electronics and Information Technology (DEIT) had taken up an initiative in February 2002 for setting up of Community Information Centres (CICs) in the hilly, far-flung rural areas of the country to bring the benefits of ICT (Information and Communication Technologies) to the people for socio-economic development of these regions and to alleviate the digital divide between urban and non-urban areas. The initiative which was as a follow up to the special package announced by PM (Prime Minister) to the North Eastern States has been extended to Jammu & Kashmir (J&K) in February 2004, Andaman & Nicobar Islands, and Lakshadweep Islands in February 2005. These CICs are a citizen interface for IT enabled e-Government services and training. During 2006, CIC at Nathula Pass in Sikkim near the Indo-China border has been commissioned which is the highest internet kiosk. 124 CICs in Jammu & Kashmir, 38 CICs in Andaman & Nicobar Islands, and 30 CICs in Lakshadweep Islands have been made operational.

Achievements of CIC: The biggest achievement of the CICs was that they made ICT penetration possible in rural area of North East. Thanks to the CICs, a large section of the rural population is now aware of the potential of ICT. They have also been trained in the basics of accessing and utilizing ICT services and facilities. CICs organized basic ICT orientation training programmes, Functioned as rural Internet cafes, as rural DTP centres, developed elementary web sites. Government officials at the block administration level have also been exposed to the potentials of ICT to a great extent due to the services provided by the CICs. The CICs have performed the crucial role of orienting the people towards ICT, and preparing the ground for more sophisticated ICT initiatives in the future.

Interim provision has been made for financial assistance up to 2 years beyond February 2007 for management of existing 555 CICs of North-Eastern States as a gap bridging arrangement to facilitate their merger into Community Service Centres (CSCs) under the NeGP (National e-

Governance Plan). The financial assistance would be available to State Governments agreeing for integration of CICs with CSC Scheme and submit the RFP for CSC Scheme.

2. **Common Service Centres (CSCs):** Government of India had launched the Common Service Centre (CSC) Project for providing sustainable digital access to rural communities and for giving the necessary impetus for upliftment of the rural community. The project is a strategic cornerstone of the National e-Governance Plan (NeGP), approved by the Government in May 2006, as part of its commitment in the National Common Minimum Programme to introduce e-governance on a massive scale. The objective is to develop a platform that can enable Government, private and social sector organizations to align their social and commercial goals for the benefit of the rural population in the remotest corners of the country. The CSC Scheme is being implemented in a public private partnership (PPP) framework with a focus on rural entrepreneurship & market mechanisms. The CSCs has been set up by private entities called as Service Centre Agencies (SCA), who are appointed by the State Governments. The CSCs are run by the Village Level Entrepreneurs (VLEs) who are appointed by the SCAs to run and manage the CSCs at pre-defined locations. To ensure availability of connectivity to all CSCs, department of Information Technology is supporting BSNL to provide Wireline and WiMax based broadband connectivity to all CSCs. CSCs are popular with different names in different state and union territories like E-mitra in urban areas of Rajasthan, E- Sampark in Chandigarh, Choice in Chhattisgarh, Jeevan in Delhi, E-Disha in Haryana, E- Lokmitra in Hinachal Pradesh, Khidmat Center in Jammu & Kashmir, Nagri Suvidha Kendra in Madhya Pradesh, Mahaseva Kendra in Maharashtra, Rainbow in Meghalaya, Jan Seva Kendra in Uttar Pradesh and so on.

   Progress of the CSC Scheme: As on 31st March 2014, 1,33,847 Common Services Centers have been set up across the country, covering 26 States and 05 UTs out of the targeted 100,000 centers to be set up one per every six villages. It is also now proposed to reposition CSCs as a network of Bharat Nirman CSCs to be set up in each Gram Panchayat as a delivery point of public services for rural areas. Accordingly, additional 150,000 Bharat Nirman CSCs are proposed to set up for the uncovered Panchayats.

Services offered by CSCs

Government to Citizen (G2C) Services: Government to citizen (G2C) services refers to online communication between Government and consumer/ individuals. It is an online interaction between local, central Government and private individuals which helps in making public services and information accessible to all. CSCs are brought in to increase the transparency in transactions at grass roots level. It has been seen that the transaction cost for services have also come down drastically and service delivery through CSCs has given citizens easy access to many services such as online form filling, bill payments, municipal certificates etc, or complex applications like distance education, e-learning, and telemedicine. Following are the key G2C services delivered through CSCs.

Services	Description
Agriculture	Farmer /Labour / Agriculturist Certificate, Crop assistance Certificate, Farmer Registration for KCC, Agricultural Family member Certificate, Appl. For incentive to green fodder cultivation, cultivation practices, land records, etc.
Ration Card	Issuance of new / increase or decrease of Units, Addition or deletion of name / renewal of ration card.
Electoral Roll	Application for Inclusion / Correction/ Cancellation / Transposition of names.
Poultry / Fisheries	Appl. Financial Assistance to Purchase cows/ buffaloes, Registration / renewal of Fishing Nets, Application for Piggery/Poultry development Scheme, Application for Modern dairy / Subsidy to purchase Milking machine Scheme.
Caste / Minority	Application for issue of SC /ST/ OBC Caste Certificate, Appl. for scholarship/ stipend to SC/OBC students, Appl. for Book Bank Scheme, Appl. for Kanya dhan for SC.
Driving License	Appl. for new / duplicate / renewal, Appl. for issue of Learner's License.
Income Certificate	Appl. for Issue of certificate, Non creamy layer certificate
Data entry / Digital Print / MIS entry	NREGA MIS entries / other digitization works, Muster roll Entry of MGNREGA, Cattle Census data digitization.
Birth / Death Certificate	Issue of birth / death certificate.
Other Certificates	Affidavits, Domicile Certificate, Bonafide, Widow, Solvency, Character, disability / senior citizen card, Non Tenancy, No Govt. Job, re-marriage, residence.
Public Grievances	Online filing of grievances.
Utility Bill payment	Electricity bill, Landline bill, Water bill.
Arms License	Application for Renewal of arm license.

(Contd.)

Services	Description
Kerosene / Retail Foodstuff License	Application for new or renewal of hawker for Kerosene Retailing / food stuff license.
PAN Card	Application for PAN card making.
Sale of Stamps	Sale of Stamps & Special adhesive stamps/PDS forms etc.
Taxes	Returns and renewal of registration for Entertainment/Entry / Luxury/Tax, Payment of VAT/VATCST/CST, Filing of returns for COT/CST/TDA.
Pension/Incentives/Grants	Application for Financial Assistance under Prime Minister's Employment Generation/Incentives for Women Entrepreneurs, Grant-In-Aid Scheme to provide Financial Assistance to the Mahila Mandals, Grants and financial assistance to professional groups and institutions/cultural organizations, Indira Gandhi National old Age Pension, Handicapped Pension/Pension/Widow.
Employment Exchange	Issue of New card/registration, issue of duplicate/renewal of card.
Forest/Environmental	Fencing registration, Application for intimation of Felled Trees, Application for relief under natural calamity
Land Records	Issuance of copy/Land holding/Land less certificate, Land Revenue Records, Transfer/Mutation of land ownership.

Business to Citizen (B2C) Services: Following Business to Citizen (B2C) services are delivered through CSCs. Agriculture services (Agri Information / Agri Input Sales / Agri Extension Services), Financial and Banking Services (Opening of Bank Accounts, Sourcing Loan Applications, Formation of SHG groups, deposits and disbursal of cash), Insurance services (Online EMIs payments), Education/ e-learning, Telecom services (Sale / recharge / Bill Collection), Travel (Train / Bus / Air ticket bookings), Entertainment Services (Sale and recharge of dish TV connections), Media and Advertising, Department of Posts (Extending Postal Services / rural Postal Insurance), Digital Photos / CD writing /Typing, Web surfing / Email/Chats, DTP / Printing and Photocopy services are made available on the payment basis etc.

3. **ISRO's Village Resource Centre:** To provide the space technology enabled services directly to the rural population, ISRO launched the Village Resource Centres (VRCs) programme in association with reputed NGOs, trusts, state and central government agencies. This project is striving to promote a need based single window delivery system for providing services in the areas of agriculture and livelihoods, education,

health, nutrition, weather, environment etc. to the rural population to empower them to face the development challenges. VRCs provides a variety of services like telemedicine, tele-education, database services on the natural resources, interactive farmers' advisories on agriculture, fisheries, weather etc. The VRC is a totally interactive V-SAT (Very Small Aperture Terminal) based network. Implementation of the project is being done jointly by ISRO and the associating agency. The associating agency is expected to set up the required infrastructure and required manpower to manage the VRCs. ISRO provides the satellite connectivity for communication, tele-education, telemedicine and required equipment. The monitoring of the progress of the project and also the quality of the technical input/ output etc. will be evaluated time to time and based on this proper corrective measures will be taken up. Total of 473 VRCs have been set up in 22 States/ Union Territories, namely Andhra Pradesh, Assam, Bihar, Delhi, Gujarat, Himachal Pradesh, Jharkhand, Karnataka, Kerala, Madhya Pradesh, Maharashtra, Meghalaya, Nagaland, Orissa, Puducherry, Rajasthan, Sikkim, Tamil Nadu, Uttarakhand, Uttar Pradesh, West Bengal and A & N islands. The VRCs are connected to knowledge or expert centres like Agricultural Universities, Skill Development Institutes and Hospitals. Over 6500 programmes have been conducted by the VRCs in the areas of, Agriculture, Horticulture, Fisheries, Live stock, water resources, Health care, Awareness programmes, Women empowerment, Supplementary education, Computer literacy, Micro credit, Micro finance, Skill development, vocational training for livelihood support etc. Over five lakh people have been availed VRC services so far.

Future Plans: Expanding the network requires additional satellite bandwidth, infrastructure, servers etc; ISRO is working in this direction. It is also proposed to set up an exclusive Content Server for storing and retrieving Programmes across the network. VRC programme will also become an integral part of the 'National Rural Knowledge Movement' to provide connectivity to all the villages in India, in which, it is planned to provide satellite based VRC connectivity to almost 4000 blocks (sub District level) by the end of 12th Five year plan.

4. **Dairy Information Services Kiosk (DISK):** DISK was initiated by the National Dairy Development Board (NDDB). DISK and the dairy information portal were conceived by IIM Ahmedabad. Dairy information kiosk is being piloted at two dairy cooperative societies in Kheda district of Gujarat, while the dairy information portal is being tested at one point in Gujarat. DISK builds on the use of an automated machine that is used

by about 2500 dairy cooperative societies (DCSs). The machine immediately tells farmers the quality and quantity of their milk, while generating price information on the procured milk. The automated machine is being implemented in 600 locations in Kheda district. The milk co-operative movement has made India the largest producer of milk in the world and spans over 70,000 villages in 200 districts of India. The unit of operation is a village level co-operative where men and women deposit their milk twice a day and get paid in return. A Personal Computer (PC) kept at the village co-operative is hooked up with the automated machines and is used for analysis. DISK is an exercise in upgrading this PC into a kind of management information system. So each of these PCs is equipped with data analysis and decision support capabilities vis-a-vis improving productivity and yield of milch cattle and in improving milk collection for the co-operative.

In addition these PCs are connected by the internet to a dairy portal that has information on dairying and best practices on milch cattle. The place where the PC is kept also doubles up as a communication centre with fax, e-mail and telephony, e-banking facilities. Government forms can also be downloaded from the internet. The portal is equipped with intranet facilities to meet the official and transaction needs of the co-operative structure. This milk testing system goes by the name of Akashganga or Milky Way.

Services offered Dairy Kiosks

a. DISK facilitates the creation and maintenance of databases of cooperative society members in the regional language. This includes information about members, their cattle, various service transactions (artificial insemination, cattle feed etc.) and the daily milk transactions.

b. Data analysis and decision support to help a rural milk collection society in improving its performance i.e. enhancing milk collection.

c. Data analysis to improve productivity of milch cattle.

d. Information on dairying, government and private sector agency services, feedback on quality of service.

e. Access to multimedia database on innovations captured by Society for Research Initiatives for Sustainable Technologies and Institution (SRISTI) on agricultural practices, medicinal plants, home remedies, tools and implements.

f. E-mail, fax and Net telephony.

g. Net banking and ATMs which will enable the milk societies to credit payments directly to sellers' bank accounts.

h. Download government forms; place orders for agricultural inputs.

i. Intranet facilities for other services.

5. **MSSRF's Village Knowledge Centres:** M S Swaminathan Research Foundation (MSSRF) is a non-profit research organization and was established in 1988. MSSRF has all along been developing and following a pro-nature, pro-poor, pro-women and pro-sustainable on-farm and non-farm livelihoods through appropriate ecotechnology and knowledge empowerment. M.S. Swaminathan Research Foundation started an experimental project in Pondicherry in 1998 with the objective to use ICT for poverty alleviation in rural areas. By December 2004 it had established Village Knowledge Centres (VKCs) in 12 villages in Pondicherry (initially there were seven), based on a 'Hubs and Spokes' model. The technological hardware was originally a makeshift modem and radio setup, but now new commercial wireless networking equipment gives villagers high-speed wireless telephone and Internet access. The current phase is utilizing and assessing Wi-Fi, 2G, mobile technology, geographical positioning system (GPS) for fisher folk, VoIP for low cost long distance voice communication, and RailTel. There is solar power backup, and video conferencing facilities. MSSRF recommends that each VKC should have 2-3 computers, a web camera, phone, printer and notice board.

   MSSRF has been scaled up its ICT work, which now comes under the umbrella of the Jamsetji Tata National Virtual Academy for Rural Prosperity (NVA). There are now a further four block level hubs in Tamil Nadu, each in a different district; and there is also a state-level hub in Chennai. The state level hub, located at MSSRF, is the knowledge resource that creates and maintains websites and databases for the local hubs in close collaboration with national and international agencies. "It is an information system that establishes lab-to-lab, lab-to-land, and land-to-land linkages" (MSSRF Annual Report, 2004-2005).

   The NVA aims to provide information and knowledge related to drought, climate management, augmentation of water, maximizing crop yield and markets. It provides the services of tele-education, tele-medicine, online decision support, interactive farmers' advisory services, tele-fishery, weather services and water management.

6. **Rural Internet Kiosks Gyandoot:** Gyandoot was initiated by District Collector of Dhar in Madhya Pradesh as rural e-government project. Dhar district having a population of 1.7 million; where 60 percent live below the poverty line. Gyandoot project was launched on January 1, 2000 with the installation of a low cost rural Intranet covering 20 village information kiosks in five Blocks of the district; later, extending to 11 more kiosks. The goal of the Gyandoot project has been to establish community-owned, technologically innovative and sustainable information kiosks in a poverty-stricken, tribal dominated rural area of Madhya Pradesh. Gyandoot is being managed by a society called 'Gyandoot Samiti'. Its objectives were to provide useful information to people in rural areas, and to act as an interface between the district government and ordinary people. For establishing the kiosks, villages that function as block headquarters were chosen where computers were installed and connected through an intranet network. This intranet has been named Gyandoot and the information kiosks are called Soochanalayas. Gyandoot' is recognized as a breakthrough in e-government, demonstrating a paradigm shift which gives marginalized tribal citizens their first ever chance to access knowledge, with minimum investment. From the Soochanalaya, user charge based services are given to the masses and at the same time the information technology related developmental needs of government departments and Panchayats are met free of cost.

   Hardware: Information kiosks have dial-up connectivity through local exchanges on optical fiber or ultra high frequency (UHF) links. The server is a remote access server housed in the computer room in the district Panchayat. Gyandoot has provided both e-governance and e-commerce services.

The services offered by Gyandoot Kiosks

a. **Agriculture Produce Auction Centres Rates:** Prevailing rates of prominent crops at the local and other recognized auction centres around the country are available online for a nominal charge.

b. **Copies of Land Records:** Documents relating to land records including khasra (record of rights) are provided at the spot on charge basis.

c. **Online Registration of Applications:** Villagers had to make several visits to the local revenue court to file applications for obtaining income, caste, and domicile certificates. Gyandoot have solved this problem.

d. **Online Public Grievance Redress:** A complaint can be filed and a reply received within 7 days.

e. **Village Auction Site:** It makes auction facilities available to farmers and villagers for land, agricultural machinery, equipment, and other durable commodities. One can put one's commodity on sale for a charge of Rs. 25/- for three months.

f. **Transparency in Government:** Villagers could step on the same platform and understand the functioning of the government better.

g. **Other Services:** Other services offered at the kiosks include online matrimonial advertisements, information regarding government programs, a forum for school children to ask questions, ask an expert, e-mail (free for information on child labour, child marriage, illegal possession of land belonging to Scheduled Tribes, etc.). Some kiosks also have added photocopy machines, STD, PCO, and horoscope services.

7. **Warana Wired Village:** In the Warana Wired Village Project covering 70 villages in Maharashtra the existing cooperative structure has been used with state of the art infrastructure to allow Internet access to existing cooperative societies. The aim is to provide information to villagers by establishing networked booths (Information kiosks) in the villages. Information kiosks in this sugarcane growing region have computers that are linked to a central network that provides farmers access to essential pieces of information such as the ideal time for planting and harvesting sugarcane, the current market rates of their produce, and payments made by the factories. The computer network has put an end to a major reason for anxiety at harvest time. Any delay in harvesting reduces its sugar content and, consequently, weight. Farmers are paid according to the crop's weight. The computer network provides each farmer with a share code. By punching the code into the system, the farmer gets details such as when the crop was planted and when it is due for harvesting. This gives the farmer sufficient time to organise workers to cut and transport the sugarcane. The network also gives details of farmer's transactions with the local sugar and milk cooperatives and helps them compare sugarcane prices in different parts of the country. The computer kiosk has made several tasks easy and less tedious for the sugarcane farmers. For instance, after sugarcane was weighed at the factory it took four days for farmers to know how much money they would get. Now, within two hours of the crop reaching the factory, farmers know how much they will be paid. The computers at the sugar factory's

weigh bridge feed the crop's weight into the farmer's file through his share code. A receipt is issued to the farmer or the transporter. The farmer can check his payment status at the computer booth (Kiosk). The booths are located at the milk collection centres in the villages. The sugar factory pays Rs.150 every month for the use of the 12 foot x 12 foot room that houses the computer. The service is free for farmers. Whenever farmers send sugarcane to the factory they go to the computer booth once in two days to check the prices and the details of his transactions with the cooperative. Farmers trust the system and know they cannot get cheated. High speed VSATs (very small aperture terminals) connected to the National Informatics Centre (NIC) in Pune and an electronic telephone exchange form the basic technology frame of the wired village project. The VSATs provide Internet access and the electronic telephone exchange provides dial-up facilities to the central hub located at the office of the Warana sugar cooperative. For the network the NIC has developed a software package based on an accounting programme. The programme supports the local language which is essential for the project.

8. **Drishtee's Kiosks:** Drishtee describes itself as an India based platform for rural networking that provides IT enabled services to rural and semi urban populations. Drishtee facilitates the creation of a rural networking infrastructure. Through a network of village kiosks, Drishtee and the local entrepreneurs deliver ICT based services to the rural population. So far it has licensed 309 kiosks. Drishtee has developed its own proprietary software platform: other service providers can use the existing network by plugging their services into the Drishtee software. Hardware package includes a computer, modem, printer and four hour battery. Almost all kiosks have dial-up connections. Drishtee is active in six districts in six states: Sirsa in Haryana, Jaipur in Rajasthan, Dewas in Madhya Pradesh, Trichy in Tamil Nadu, Sonitpur in Assam and Madhubani in Bihar.

   Services offered by Kiosks: Users are provided with enhanced access to e-Governance, education, health, insurance and local services. Drishtee is providing localised content for educational and health services. In the case of e-Government services the aim is to provide rural citizens with doorstep access to several services for which they previously spent a lot of time and money to get the job.

9. **UBI's Village Knowledge Centres:** As part of commitment to the nation, Union Bank of India has opened more than 200 Village Knowledge Centres (VKC) throughout the country for capacity building of rural people. First VKC was set up in May, 2005. For this innovative initiative

the bank has been awarded the prestigious "Skoch Challenger Award" in March, 2007. For all round socio economic development of villages, 101 villages across the country have been adopted through selected Village Knowledge Centres under "Union Adarsh Gram" project. Currently 201 VKCs are operational. Village Knowledge Centre serves as information dissemination centre providing instant access to farmers to latest information or knowledge available in the field of agriculture, starting from crop production to marketing.

Role of VKC Incharge

a) Impart knowledge on new developments, improved methods of cultivation technologies in the field of agriculture.

b) Dissemination of weather data and agro climatic conditions, latest information on prices of agriculture produce to farmers.

c) Imparting knowledge on diversification of Agriculture and developing at least 2 farm-based Enterprises in a village per year.

d) Bring in positive change in the livelihood of at least 300 farmers in a year in the command area of a VKC by increased income levels and higher standards of living.

e) Form at least 5 Self Help Groups per village in the command area of VKC.

f) Provide constant support to the SHGs to ensure that at least 75 percent sustain beyond 6 months to become eligible for bank linkage.

g) Enable the groups to graduate economically for taking up income generating activities and increasing their credit absorption capacity.

h) Form at least 1-2 Farmers Clubs per village in the command area of VKC.

i) Create awareness in the command area villages about various government sponsored schemes to take benefit of subsidy/grants.

j) Coordinate with Govt. departments to ensure flow of funds to the villages allocated under various developmental schemes.

k) Ensure that every family in the command area villages has at least one active bank account.

l) Provide financial counselling and inform about various schemes /products of our Bank to at least 5 people in a day.

m) Special attention and counselling for the vulnerable people who are heavily indebted to informal sources of finance and defaulter to Banks/ FIs to come out of debt trap. Work as a relationship manager between the Bank and the farmers.

UBI's VKC Portal: To support the activities of VKC, UBI has launched "Village Knowledge Center Portal" in Bank's Intranet. For accessing "VKC Portal" (http://172.31.14.205/vkc), go to UBINET (http://172.31.14.205/ubinet) and click "VKC Portal" under Useful Links. VKC Portal would provide all the information related to VKC and links to different website addresses that VKC In-charge need to access to provide updated information to the farmers. VKC Portal can be accessed for the following information:

- Past 24 hrs Weather Report and Today's Weather Forecast.
- Weather Forecast for the Next 5 Days.
- Weekly Crop Advisory.
- Kisan Call Centre.
- Market Prices of Agri Commodities.
- Latest Technologies in the field of Agriculture.
- Animal Husbandry.
- Farm-based Enterprises.
- Model Bankable Projects.
- Govt. Schemes Related to Agriculture.
- Websites of State Agriculture Departments.
- Websites of Agricultural Institutes.

10. **ISAP- Community Technology Learning Centres (CTLCS):** ISAP with support of Microsoft - Unlimited Potential Programme has established 'Community Technology Learning Centres (CTLCs)' in remote villages of Maharashtra to provide IT training to 45,000 farmers and unemployed youth. Under two-year programme, ISAP would be setting up 250 CTLCs at village level for imparting IT training to rural community and increase their income earning potential. ISAP is working on online weekly price monitoring system of herbal & medicinal plants with the funding support of National Medicinal Plant Board. ISAP gathers and manages authentic data about the weekly price and demand for 101 medicinal plants from 50-marketing centers in different states of the

country. These data are weekly upgraded on the basis of prices and quantity offered for different medicinal plants.

11. **ITC's e-Choupal:** E-Choupal is a Hindi word which means "village meeting place". Market is a meeting place where vendors and customers come together to do transactions. e-choupal is a virtual market place where farmers can transact directly with a processor and can realize better price for their produce. e-choupal has the advantages of the market but spans very large varieties of vendors and customers. Geographical distances do not restrict participation in the e-choupal. The main disadvantage of conventional market is that information asymmetry is inherent in the market whereas, e-choupal provides transparent transactions. This enables the participation of smaller as well as larger players. Elimination of some layers of intermediaries allows for larger share of profits to reach the lower end of value chain. The main attractiveness of e-choupal is that it can be used for connecting large producers/ small producers and small users/ large users, thereby eliminating the need for hierarchy of brokers. Internet is used as a low transaction cost backbone for communication. Physical delivery of produce to the processor is still done through the existing intermediaries. e-choupal does not attempt total elimination of intermediaries, as intermediaries are indispensable in economy like India where intermediaries are adding value to the every step of value chain at a low cost. Intermediaries have the expertise in storage, transportation, quality assessment and counter party risk reduction, which are difficult to replicate. e-Choupal provides farmers with all the market information and this helps them to become market oriented. In e-choupal intermediaries are leveraged but they are disintermediated from the market information flowing to the farmers.

    The project e-Choupal is ITC's unique click & mortar initiative. e-Choupal is an ICT platform for carrying out trade at a number of locations. In this, ITC sets up a back-up physical service support at the village level, called e-Choupal, through Sanchalak: a lead farmer, who acts as the interface between computer and the farmer. ITC accumulates information regarding weather, modern farming practices, and market prices from sources like Meteorological Department, Agriculture universities; mandis (regional market) etc., and upload all information on to e-Choupal website. All information is customized according to local farmer's requirements and provided into the local language through computer set up by ITC in Sanchalak's house. Sanchalak access this information and facilitates its dissemination to farmers. Information regarding weather and scientific farming helps farmers to select the right crop and improve the productivity of their farms. Availability of market information helps farmers to become market oriented. They know what price ITC is quoting and the price prevalent in the local market, thereby helping better price realization for farmers. If farmer decides to sell to ITC, Sanchalak works as the aggregator of small farmers produce to sell them to ITC. Sanchalak also aggregates farmers input purchase orders for various items like seeds, pesticides and places them directly with the suppliers through internet and facilitate supply of high quality farm inputs as well as purchase of farm produce at farmer's doorstep with the help of intermediaries. It can be deciphered that e-Choupal has added critical value to the existing supply chain through innovative application of information technology. ITC has successfully reached the vastly scattered farming villages of India and facilitated the smooth flow of rich information to them using ICTs. Power of information is working as the catalyst of transformation of the life of farmers by helping them to get improved yields from their farms and better price realization. Usually it is hard to maintain the expanding scale of reach and richness into the products or services but e-Choupal is achieving it easily and we find that overall value chain has been shifted horizontally on the scale of reach and richness. e-Choupal facilitates flow of information and knowledge, and supports market transactions on line as under.

- It transmits Information (weather, prices, news),
- It transfers Knowledge (farm management, risk management)
- It facilitates sales of Farm Inputs (screened for quality) and
- It offers the choice of an alternative Output-marketing channel (convenience, lower transaction costs) to the farmer right at his doorstep.
- It is an interlocking network of partnerships (ITC + Meteorology Dept. + Agri. Universities + Input Companies + Sanyojaks, the erstwhile Commission Agents) bringing the "best in class" in information, knowledge and inputs.

Launched in June 2000, 'e-Choupal', has already become the largest initiative among all Internet-based interventions in rural India. 'e-Choupal' services today reach out to over 04 million farmers growing a range of crops - soybean, coffee, wheat, rice, pulses etc. in over 40,000 villages through 6500 kiosks across 10 states (Madhya Pradesh, Haryana, Uttarakhand, Karnataka, Andhra Pradesh, Uttar Pradesh, Rajasthan, Maharashtra, Kerela and Tamil Nadu).

This enthusiastic response from farmers has encouraged ITC to plan for the extension of the 'e-Choupal' initiative to all together 15 states across India over the next few years. ITC has planned to channelise other services related to micro-credit, health and education through the same 'e-Choupal' infrastructure.

12. **Project Bhoomi:** The Department of Revenue in Karnataka State has computerized 20 million records of land ownership of 6.7 million farmers in the State. Previously, farmers had to seek out the village accountant to get a copy of Record of Rights, Tenancy and Crops (RTC) – a document needed for many tasks such as obtaining bank loans. There were delays and harassment. Bribes had to be paid. Today, for a fee of Rs. 15, a printed copy of the RTC can be obtained online at computerized land record kiosks (Bhoomi centers) in 177 taluk offices. This system works with the software called "BHOOMI" designed fully in-house by National Informatics Center, Bangalore. The Department of Information Technology, Govt. of India has embarked upon a major programme to rollout Land Records Computerization in several States of the country. Presently most of the states have comprised their land record. Farmers can get their land records easily from e-kiosks by paying nominal fee.

Grameen Gyan Abhiyan: Every Village a Knowledge Centre

During 2004, M S Swaminathan Research Foundation initiated steps to extend its model of ICT-enabled development activities to different parts of the country in the consortium mode involving several partners (IDRC, Tata Trusts, DIT, IIT-Chennai, ISRO, NASSCOM, Microsoft, INCOIS, etc.). Since then, the list of partners has been growing day by day. Thus, Mission 2007 has triggered a national movement for bridging the urban-rural digital divide and for ensuring knowledge connectivity in areas relevant to the day-to-day life and livelihood of rural families. The then plan was to provide knowledge centres in all the 600,000 plus villages of India by the 60th anniversary of India's independence that fell on the 15th August 2007. Since early August 2007 this network is referred to as "Grameen Gyan Abhiyan (Rural Knowledge Movement)" as suggested by several partners. They suggested that from now onwards the concept should become movement. Some of the GGA partners are testing different technologies and developing applications for ICT-based rural centres. Academicians, Corporate Sectors and Policy Makers could use this platform for knowledge and technological empowerment.

CHAPTER 11

Status and Prospects of ICT Application in Agriculture

ICTs are emerging as very important tools for agricultural extension and it is now essential for every extension worker to have working knowledge of ICT tools and devices like computers, internet, mobile phone, kisan call center, information kiosks, e-mails, expert systems, touch screen systems, mobile telephony and World Wide Web. Keeping in mind wide scope of ICT application in agriculture, the entire study entitled "Status and prospects of ICT application in agriculture through extension functionaries" was undertaken with following specific objectives.

i) To assess the present status of knowledge of extension functionaries about selected information and communication technologies.

ii) To determine the extent of application of defined information and communication technologies in agriculture by the respondents.

iii) To study the attitude of extension functionaries towards application of information and communication technologies in agriculture.

iv) To assess the training needs of extension functionaries regarding application of information and communication technologies in agriculture.

v) To identify the constraints perceived by the extension functionaries in application of information and communication technologies in agriculture.

vi) To explore the possibilities of information and communication technologies application in agriculture for better prospects as viewed by the extension functionaries.

Research Methodology

The study was conducted in Udaipur district of Rajasthan. Udaipur district was selected purposively on the basis of maximum number of extension functionaries are working in the state department of agriculture and a large number of NGOs in the district. For the purpose of study government and non-government organisations actively involved in the transfer of agricultural technology were selected. The transfer of agricultural technology in the district is being carried out by state department of agriculture extension so the extension unit of state department of agriculture was selected purposively as Government Organisation (GO). To select the NGOs a complete list of all the NGOs using ICTs in the transfer of agriculture technology among the farming community was prepared. From the list so confirmed eight NGOs with more use of ICTs were selected.

To select a sample of respondents from the GO, a preliminary list of all those extension functionaries who are engaged in transfer of agriculture technology work and using ICTs was prepared. From the list, so prepared 80 extension functionaries were selected on the basis of random sampling technique. Likewise, for the selection of respondents from NGOs, organisation wise list of all those field level functionaries who are engaged in transfer of agriculture technology work was prepared. From the list 80 extension functionaries from selected NGOs was taken on the basis of proportionate sampling procedure. Consequently, a total of 160 respondents (80 from GO & 80 from NGOs) were included in the sample of study.

For the purpose of study five ICT tools namely computer, internet, mobile phone, kisan call centers and information kiosks were taken for the study. It was observed that these identified tools are commonly being used by both GOs and NGOs for transfer of technology in agriculture. A comprehensive interview schedule was developed considering the specific objectives formulated under the study. Thereafter, data were collected from the selected respondents by employing personal interview technique. The interviews were conducted at extension personnel's homes as well as office in the month of April to June 2012. About 45-60 minutes were spent to interview each respondent. The hypotheses were formulated and appropriate statistical tests were used to arrive at specific conclusions.

Major Findings of the Study

The important findings, which emerged from the study, have been presented under the following heads:

Background Information of the Respondents

i) The study indicated that out of 160 respondents, majority of extension functionaries (60.62%) belonged to the age group of 31 to 45 years. Whereas, (22.50%) extension functionaries were reported from age group of up to 30 years and remaining 27 respondents (16.88%) were found in the age group of 46 to 60 years.

ii) It is evident that 32.50 per cent extension functionaries were high school passed, while 35.62 per cent extension functionaries were graduates. A considerable number (27.50 %) of the respondents were reported post graduate in various disciplines and only 4.38 per cent of the extension functionaries were educated up to doctoral level.

iii) The results visualized that out of total 160 respondents 38.75 per cent respondents had their income one to three lacs per annum and 41.88 per cent extension functionaries had their income three to five lacs per year, while remaining 19.37 per cent of the total respondents earned above five lacs per annum. A comparative view of annual income of extension functionaries of GO and NGOs highlights that majority of the NGOs personnel (66.25%) were earning Rs. 01 to 03 lacs per year while, majority of GO personnel (53.75%) were reported in the medium income level of Rs. 03 to 05 lacs in the study area.

iv) The study reported that 31.88 per cent of the extension functionaries did not receive any trainings on ICTs while, 36.25 and 31.88 per cent respondents received short and long duration training respectively. It was very interesting fact that 41.25 per cent of GO functionaries and 22.50 per cent of NGOs functionaries did not receive any kind of training on ICTs whereas, 40 per cent of GO functionaries and 32.50 per cent of NGOs functionaries got short duration training on ICTs. It was also noted that 18.75 per cent GO respondents and 45 per cent NGOs respondents received skill based long duration training which enabled them to work with various ICT tools.

v) It is evident from the study that 35.62 per cent extension functionaries possessed low computer proficiency, while 30.63 per cent of the extension functionaries had medium level of proficiency in operating the computers. Remaining 33.75 per cent of respondents were observed in the high computer proficiency group.

vi) The study showed that out of total 160 respondents, 38.75 per cent of extension personnel possessed high level of experience about handling ICT devices, whereas, similar per cent of respondents (38.75%) had medium level of experience about ICT devices. Remaining 22.50 per cent of extension personnel were categorized into low level of experience about handling of ICT device.

vii) It is evident from the study that 36.25 per cent extension functionaries had low level of accessibility of computer and internet services whereas, 36.25 per cent extension functionaries had high level of accessibility of computer and internet services followed by 27.50 per cent respondents who had availability of computer and internet facility at medium level.

Knowledge of the Extension Personnel about ICT

i) The study revealed that majority of the total respondents 71.88 per cent had medium level of knowledge about information and communication technologies whereas, 20.65 per cent respondents were observed in low level knowledge group. It was further noted that only 7.50 per cent of the total respondents had high level of knowledge regarding ICTs in the study area. The results further indicated that 57.50 per cent respondents of GO and 86.25 per cent NGOs respondents had medium level knowledge about ICTs. On the other hand 35.00 per cent of GO respondents and only 6.25 per cent of NGOs respondents belonged to low knowledge group. A considerable number of GO respondents in this group clearly show that they have low level of knowledge about ICTs. However, an equal number of respondents from both GO and NGOs *i.e.* 7.50 per cent were reported in the high level of knowledge regarding ICTs.

ii) The results of the study revealed that the extent of knowledge of GO personnel was from 7.50 to 98.75 per cent, whereas in case of NGOs respondents the extent of knowledge was observed to be 43.75 to 100.00 per cent in all aspect of computer. It means that extension functionaries of NGOs possessed more knowledge than extension functionaries of GO about all the aspects of computer technology.

iii) It was found that the extent of knowledge among GO respondents was 8.75 to 88.75 per cent, whereas in case of NGOs respondents extent of knowledge was observed to be from 43.00 to 100 per cent in all the aspects of Internet technology. It inferred that NGOs personnel had comparatively more knowledge than GO personnel about various aspects of internet technology.

iv) It can be concluded that the extent of knowledge of GO respondents was 27.50 to 100.00 per cent, whereas, extent of knowledge of NGOs respondents was reported to be 46.87 to 100.00 per cent in all aspects of mobile phone technology. Further it was inferred that NGOs personnel had greater knowledge than GO personnel except some aspects of cell phone technology.

v) It was found that extent of knowledge among extension personnel of GO was 5.00 to 100 MPS, while in case of NGOs respondents it was 21.25 to 88.00 MPS about all aspects of kisan call center. The data also inferred that respondents of GO had greater knowledge than NGOs respondents about various aspects of kisan call center.

vi) It was reported that the extent of knowledge in GO personnel was from 17.50 to 83.75 MPS, while in case of NGOs personnel the extent of knowledge was observed to be from 20.00 to 83.25 MPS about all aspects of information kiosk.

vii) Results of the study revealed that there was significant difference in knowledge between GO and NGOs extension functionaries about information and communication technology. This difference in the knowledge of extension personnel might be due to the reason that NGOs respondents have more availability and accessibility of ICT tools in their office as well as home and using these ICT tools in daily life along with professional life.

Application of ICT in Agriculture by the Extension Personnel

i) The findings revealed that majority (56.25%) of the total respondents were found in the medium group of ICT application, whereas, 29.37 per cent respondents were in low group of ICT use and remaining 14.38 per cent extension functionaries to be observed in the high group of ICT application. It can be further concluded that more than 71.00 per cent of respondents from NGOs were to be observed either from medium or high ICT application group whereas, about 68 per cent GO personnel were found to use ICT from lower to medium level in the study area.

ii) The findings indicated that the extent of application of computer among GO respondents was 32.18 to 44.37 MPS, while in case of NGOs respondents the extent of computer application was 35.46 to 52.18 MPS in all the aspects of the computer technology. Further, it was noted that NGOs personnel had more extent of use than GO personnel. However, there is still exists a wide gap in the application of computer technology in agriculture.

iii) It was noted that the extent of use of internet among GO respondents was 25.00 to 44.68 per cent, while in case of NGOs respondents it was

30.25 to 58.75 per cent about various aspects of the internet application in agriculture. Further, it was noted that NGOs personnel had greater extent of use of internet than GO personnel.

iv) Findings indicated that the extent of use of mobile phone among GO respondents was 5.00 to 74.25 per cent, while in case of NGOs respondents the extent of use of mobile phone was 16.44 to 79.68 per cent about all aspects of the mobile phone technology. Further, it was concluded that extent of use of mobile phone by NGOs was more than GO personnel in agriculture.

v) The study revealed that the extent of use of Kisan Call Center by GO respondents was 27.50 to 47.50 per cent, while in case of NGOs respondents the extent of use of KCC was 27.81 to 39.37 per cent about all the aspects. Further, it was noted that extent of use of KCC among GO and NGOs personnel was more or less similar pattern.

vi) It was found that the extent of use of information kiosks by GO respondents was 25.00 to 36.87 per cent, while in case of NGOs respondents it was 25.62 to 38.75 per cent about all aspects of the information kiosks. Further, it was inferred that GO and NGOs personnel were using information kiosks more or less same.

vii) The findings revealed that there was no significant difference in the application of information and communication technologies between GO and NGOs extension functionaries. It means that extension functionaries of both the organisations were using information and communication technologies more or less similar level for extension work in the study area.

Attitude of Extension Functionaries Towards Application of Information and Communication Technology in Agriculture

i) It was found that out of total 160 respondents, 56.87 per cent had favourable attitude towards ICTs application in agriculture. Whereas, 25.63 per cent of extension personnel had most favourable attitude and 17.50 per cent respondents possessed least favourable attitude towards ICT application in agriculture. It means that majority of the respondents had positive and favourable attitude towards ICT application in agriculture.

ii) The findings indicated that most of the respondents had positive attitude towards various aspects of information and communication technology application in agriculture. It evidently proves the fact that ICTs are very beneficial to the farming communities and have power to revolutionize agriculture extension system of the country. It was further inferred that NGOs respondents possessed slightly more favourable attitude than GO respondents. Results also inferred that there was a significant correlation between GO and NGOs respondents with regards to ranks assigned to different aspects of attitude towards ICTs.

iii) The results of investigation indicated that there was no significant difference in attitude between GO and NGOs respondents regarding ICT application in agriculture. It means that both the categories of respondents had more or less similar attitude towards ICT application in agriculture.

Training Needs of Extension Functionaries About Information and Communication Technology

i) The outcome of the study revealed that majority of the extension personnel 93 (58.13%) were in the medium training need group. While, 49 respondents (30.63%) had high level of training needs about ICT application in agriculture and 18 (11.25%) respondents were observed in low training need group. Further, among the categories of the respondents 56.25 and 60.00 per cent respondents of GO and NGOs expressed medium level of training needs respectively. While, 33.75 per cent of GO respondents and 27.50 per cent of NGOs personnel fell under the category of high level of training needs about ICTs application in agriculture. The less number of respondents of GO and NGOs had low level of training needs regarding ICT application in agriculture.

ii) It was found that the extent of training needs of GO respondents was 52.92 to 80.42 per cent in all aspects of computer, while in case of NGOs respondents it was from 51.66 to 68.75 per cent. It means that GO respondents were more interested in computer training than NGOs respondents.

iii) The findings revealed that the extent of training need among GO extension functionaries was 53.33 to 87.50 per cent, whereas extent among NGOs functionaries was reported to be 46.66 to 56.25 per cent about various aspects of internet application.

iv) The outcomes of the investigation indicated that extent of training needs among the GO extension workers was 61.67 to 86.27 per cent whereas, in case of NGOs personnel it was reported to be 48.33 to 66.25 per cent which clearly indicates that GO personnel require more training about various aspects of mobile phone application in agriculture.

v) It is evident from the results that extent of training need among GO personnel was 50.83 to 85.75 per cent, whereas in case of NGOs personnel it was found 46.66 to 76.25 per cent in all aspects of kisan call center. It means that NGOs personnel needed more trainings than GO personnel about KCC.

vi) The findings shows that extent of training needs among the GO respondents was 55.83 to 78.75 MPS while, in case of NGOs personnel it was observed 57.08 to 77.92 MPS in all aspects of information kiosk.

vii) The results of the study revealed that there was significant difference in training needs of GO and NGOs extension functionaries about ICT. The mean value further indicates that GO personnel had more training needs than NGOs personnel about ICT application in agriculture.

Constraints Perceived by Extension Functionaries in Application of Information and Communication Technology in Agriculture

i) The investigation revealed that 56.25 per cent GO and 61.25 per cent NGOs extension personnel fell under the category of medium level of constraints. Likewise, 35.00 per cent GO and 22.50 per cent NGOs personnel was placed under the category of high level of constraints. However, only 8.75 per cent GO and 16.25 per cent NGOs personnel were noticed under the category of low level of constraints in the study area. It was concluded that GO extension personnel possessed more constraints than NGOs personnel in the application of ICTs in agriculture.

ii) The study indicated that lack of knowledge about computer application in agriculture, high cost of computer & devices, high charges of internet use for accessing information, poor ICT infrastructure and poor network for accessing internet at village level, non-availability of mobile phone network in the remote area, busy network of kisan call center, less number of kisan call centers, lack of agriculture information provided by information kiosk and inadequate number of information kiosks were important constraints perceived by GO and NGOs respondents in application of ICTs in agriculture.

iii) The findings revealed that there was significant difference in constraints perceived by GO and NGOs extension functionaries about ICT application in agriculture. This difference may be due to the reason that GO personnel have less accessibility to ICTs than NGOs personnel.

Possibilities of ICT Application in Agriculture for Better Prospects as Viewed by Extension Functionaries:

i) The findings indicated that computer with internet facility be made available on free of cost to extension personnel, the number of phone lines at KCC should be increased along with instant messaging, touch screen system with graphic technology be made available at information kiosks, call centers experts must be professional agriculture specialists, free mass messaging service to the agriculture offices and digital agricultural technological packages be made available in Hindi were most important possibilities viewed by the extension personnel for better prospects of ICT application in agriculture.

Salient Recommendations

On the basis of results obtained from the study and observations made during investigation, the following recommendations were made for promotion of information and communication technology application in agriculture.

1. During the investigation it was found that a large number of positions of extension functionaries are lying vacant in the state agriculture department especially in Udaipur district. At the present time 109 extension functionaries are working in the entire district, out of them many extension personnel have additional charge of other vacant posts which has widened the ratio of farmers to extension personnel pathetically. Although the appointed Village Level Workers (VLWs) are disseminating the agriculture technologies among the farmers, but they hardly accept any accountability. Looking to this situation information and communication technologies are only option in the hand of extension professionals to accelerate transfer of technology among the farmers. The findings of this study indicates that knowledge level of the extension functionaries about various ICT tools namely computer, internet, mobile phone, kisan call center and information kiosks was not enthusiastic in the study area. Therefore, it is recommended that more awareness drives be initiated by the government and other concerning agencies to increase the status of knowledge among extension personnel about various aspects of ICTs. Such awareness drive will definitely increase the knowledge of the extension personnel about information and communication technologies in the study area.

2. The result of the investigation revealed that application of ICTs in agriculture by the extension personnel was not encouraging. It was found that overall use of computer, internet, mobile phone, kisan call center

and information kiosks in agriculture was poor in the study area. To enhance the application of computer in agriculture it is recommended that computer or laptop should be provided to the extension functionaries so that they can use it for wellbeing of farming community. It is recommended that internet services should be made accessible to the extension functionaries so that they can directly access latest agriculture technologies form the websites of KVKs, SAUs, ICAR, department of agriculture and other national research centers to deliver latest information among the farmers. To increase application of mobile phone in agriculture it is recommended that mobile charges of extension personnel be made reimbursed for frequent contact with farmers. To improve use of KCC among extension workers it is advised that at least one KCC line should be provided for extension personnel where subject matter specialists should address the queries of extension personnel. Further to increase the application of information kiosks it is recommended that agriculture technological information should be made available at information kiosks on priority basis.

3. Training is an important tool, which can be effectively used to improve, update and refresh the knowledge, understanding and skills of learners. A systematically organized training programme brings about desirable changes in behaviour of the people. The findings of investigation shows that overall training needs of extension personnel regarding ICTs were higher which indicate that intensive training programmes on ICTs are required to be organized. Hence, it is recommended that skill oriented trainings on various aspects of computer, internet, mobile phone, kisan call center and information kiosks by special experts should be provided to the extension functionaries of government and non-government organizations of Udaipur district. The trained extension functionaries of GO and NGOs will certainly revitalize the extension system of the Udaipur district.

4. Indian Council of Agriculture Research has developed large number of CDs and DVDs on various agricultural technologies. These CDs are very useful to the farmers as well as extension personnel for self learning and as a training tool. Findings of the study revealed that non-availability of CD-ROMs on agriculture technologies was perceived as important constraint by extension functionaries therefore, it is recommended that CD-ROMs developed by ICAR, SAUs, and KVKs should be converted into local language and be made available to the extension personnel so that they can be utilized for agriculture development in the area.

5. Irregular supply of electricity has ever restricted application of information and communication technologies in agriculture. Hence, it is recommended that powerful UPS with long backup and power generator should be made available at least in the block level offices of extension personnel so that computers, internet and other ICT tools can be used regularly.

6. Internet is being widely used in the urban area. Most of the government, private offices and individuals have internet connectivity. Finding of the study shows that less than thirty percent of extension personnel are using internet for better service to the farmers but it was noted that understanding of extension workers about English language is poor and online agriculture content is not available in local language, therefore, it is recommended that state department of agriculture extension and KVK of Udaipur district should develop their websites in Hindi language and agriculture technological contents should be uploaded in local language on their websites. This will help to the extension functionaries for getting agricultural information through internet in Hindi and local language.

7. Internet services like broadband, Wi-Fi, 3G, EDGE (Enhanced Data Rates for Global Evolution) data cards, are costly and hardly affordable by the extension workers so the extension functionaries realized the problem of high charges of internet use for accessing information, therefore, it is recommended that internet services to the extension personnel should be provided free of cost.

8. Mobile communication technology has become the world's most common way of transmitting voice, data and services, no other technology has ever spread so faster. The mobile phone is especially important for India because it is growing fast in the country. Mobile phone is being used for transfer of various agriculture information like improved variety seeds, uses of pesticides, fertilizers, soil health, improved cultivation technologies, farm machinery, trainings, best agricultural practices, weather advisories, market prices, procurement points, monitoring, implementation, evaluation of schemes & programmmes, irrigation management, drought relief and management, livestock management etc. But mobile services are paid services which restricts its application by the extension functionaries. Non-availability of mobile supported audio-videos files on agriculture technologies is another problem faced by extension personnel. Thus, it is advised that specialized tariff plan should be provided to the extension workers and mobile supported clips on various agriculture technologies should be made available to the extension workers as well as farmers.

9. Interactive Voice Response Service (IVRS) for agriculture is as innovative ICT based idea for enhancing agriculture information services to the farmers and extension workers therefore, it is recommended that interactive voice response service 'Kisan Care' should be initiated by the agriculture department or KVKs.
10. The poor availability of internet and mobile phone network in the rural areas was perceived as important problem in the study area. Looking into wide scope of ICT application in agriculture, it is recommended that ICT infrastructures for accessing high speed internet and mobile services at village level even in the remote areas should be developed and pro ICT policy should be promoted.
11. During the investigation it was also found that KCC network was very busy and it was difficult to contact with KCC experts. It was due to high number of caller and less number of KCC line. It was also observed that more time was taken by experts to understand and diagnose the area specific problems and the solutions. Therefore, it is recommended that KCC service should be enhanced to the zonal level. It will also help to better understanding about area specific problems. It is also suggested that mobile based instant messaging service from KCC to caller should be started so that the solution of the problem can be received in the form of text message during the discussion with KCC expert.
12. Findings of the investigation revealed that numbers of information kiosk in the study area were inadequate. Non-availability of accurate and sufficient agriculture information at information kiosks was another important constraint perceived by the extension functionaries in the use of information kiosks. Hence, it is recommended that information kiosks should concentrate towards agriculture services and agriculture graduate should be preferred as kiosk operator. Further number of information kiosks in the area must be increased.

FSC
www.fsc.org
MIX
Papier | Fördert gute Waldnutzung
FSC® C083411

Zeitfracht Medien GmbH
Ferdinand-Jühlke-Straße 7
99095 Erfurt, Deutschland
produktsicherheit@kolibri360.de